Principles and Practices for a Federal Statistical Agency

FOURTH EDITION

Committee on National Statistics

Constance F. Citro, Margaret E. Martin, and
Miron L. Straf, *Editors*

Division of Behavioral and Social Sciences and Education

NATIONAL RESEARCH COUNCIL
OF THE NATIONAL ACADEMIES

THE NATIONAL ACADEMIES PRESS
Washington, D.C.
www.nap.edu

THE NATIONAL ACADEMIES PRESS 500 Fifth Street, N.W. Washington, D.C. 20001

NOTICE: The project that is the subject of this report was approved by the Governing Board of the National Research Council, whose members are drawn from the councils of the National Academy of Sciences, the National Academy of Engineering, and the Institute of Medicine. The members of the committee responsible for the report were chosen for their special competences and with regard for appropriate balance.

This study was supported by Grant No. SBR-0453930 between the National Academy of Sciences and the National Science Foundation, which provides funding from a consortium of federal agencies. Any opinions, findings, conclusions, or recommendations expressed in this publication are those of the author(s) and do not necessarily reflect the view of the organizations or agencies that provided support for this project.

Principles and practices for a federal statistical agency / Constance F. Citro, Margaret E. Martin, and Miron L. Straf, editors. ; Committee on National Statistics, Division of Behavioral and Social Sciences and Education, National Research Council. — 4th ed.
p. cm.
Includes bibliographical references.
ISBN 978-0-309-12175-0 (pbk.) -- ISBN 978-0-309-12176-7 (pdf) 1. United States—Statistical services. I. Citro, Constance F. (Constance Forbes), 1942- II. Martin, Margaret E. III. Straf, Miron L. IV. National Research Council (U.S.). Committee on National Statistics. V. National Research Council (U.S.). Division of Behavioral and Social Sciences and Education.
HA37.U55P75 2009
352.7'50973—dc22
2009003377

Additional copies of this report are available from the National Academies Press, 500 Fifth Street, N.W., Lockbox 285, Washington, D.C. 20055; (800) 624-6242 or (202) 334-3313 (in the Washington metropolitan area); Internet, http://www.nap.edu.

Printed in the United States of America

Suggested citation: National Research Council. (2009). *Principles and Practices for a Federal Statistical Agency, Fourth Edition.* Committee on National Statistics. Constance F. Citro, Margaret E. Martin, and Miron L. Straf, Editors. Division of Behavioral and Social Sciences and Education. Washington, DC: The National Academies Press.

THE NATIONAL ACADEMIES

Advisers to the Nation on Science, Engineering, and Medicine

The **National Academy of Sciences** is a private, nonprofit, self-perpetuating society of distinguished scholars engaged in scientific and engineering research, dedicated to the furtherance of science and technology and to their use for the general welfare. Upon the authority of the charter granted to it by the Congress in 1863, the Academy has a mandate that requires it to advise the federal government on scientific and technical matters. Dr. Ralph J. Cicerone is president of the National Academy of Sciences.

The **National Academy of Engineering** was established in 1964, under the charter of the National Academy of Sciences, as a parallel organization of outstanding engineers. It is autonomous in its administration and in the selection of its members, sharing with the National Academy of Sciences the responsibility for advising the federal government. The National Academy of Engineering also sponsors engineering programs aimed at meeting national needs, encourages education and research, and recognizes the superior achievements of engineers. Dr. Charles M. Vest is president of the National Academy of Engineering.

The **Institute of Medicine** was established in 1970 by the National Academy of Sciences to secure the services of eminent members of appropriate professions in the examination of policy matters pertaining to the health of the public. The Institute acts under the responsibility given to the National Academy of Sciences by its congressional charter to be an adviser to the federal government and, upon its own initiative, to identify issues of medical care, research, and education. Dr. Harvey V. Fineberg is president of the Institute of Medicine.

The **National Research Council** was organized by the National Academy of Sciences in 1916 to associate the broad community of science and technology with the Academy's purposes of furthering knowledge and advising the federal government. Functioning in accordance with general policies determined by the Academy, the Council has become the principal operating agency of both the National Academy of Sciences and the National Academy of Engineering in providing services to the government, the public, and the scientific and engineering communities. The Council is administered jointly by both Academies and the Institute of Medicine. Dr. Ralph J. Cicerone and Dr. Charles M. Vest are chair and vice chair, respectively, of the National Research Council.

www.national-academies.org

COMMITTEE ON NATIONAL STATISTICS
2008-2009

Acknowledgments

The Committee on National Statistics thanks the many people who contributed their time and expertise to the preparation of this report. We are most appreciative of their cooperation and assistance.

In expressing our gratitude to the staff, a special measure of recognition is due to Margaret Martin and Miron Straf, directors of the Committee on National Statistics in 1972-1978 and 1987-1999, respectively, who were coeditors of the first edition of this report. In preparing subsequent editions, they were joined as editors by Constance Citro. This edition benefited from the editing of Eugenia Grohman of the Division of Behavioral and Social Sciences and Education and from editing and graphic assistance by Daniel Cork of the committee staff. We also are indebted to many others who offered valuable comments and suggestions, too numerous to mention.

This report has been reviewed in draft form by individuals chosen for their diverse perspectives and technical expertise, in accordance with procedures approved by the Report Review Committee of the National Research Council (NRC). The purpose of this independent review is to provide candid and critical comments that will assist the institution in making its published report as sound as possible and to ensure that the report meets institutional standards for objectivity, evidence, and rcsponsiveness to the study charge. The review comments and draft manuscript remain confidential to protect the integrity of the deliberative process. We wish to thank the following individuals for their review of this report: Alfred Blumstein, H. John Heinz III College, Carnegie Mellon University; Daniel Kasprzyk,

Surveys and Information Services Division, Mathematica Policy Research, Inc., Washington, DC; Janet Norwood, consultant, Chevy Chase, MD; Fritz Scheuren, Vice President's Office, National Opinion Research Center, The University of Chicago; and John H. Thompson, President's Office, National Opinion Research Center, The University of Chicago.

Although the reviewers listed above have provided many constructive comments and suggestions, they were not asked to endorse the conclusions or recommendations nor did they see the final draft of the report before its release. The review of this report was overseen by John C. Bailar, III, Department of Health Studies (emeritus), The University of Chicago. Appointed by the NRC's Report Review Committee, he was responsible for making certain that an independent examination of this report was carried out in accordance with institutional procedures and that all review comments were carefully considered. Responsibility for the final content of this report rests entirely with the authoring committee and the institution.

Finally, we recognize the many federal agencies that support the Committee on National Statistics directly and through a grant from the National Science Foundation. Without their support and their commitment to improving the national statistical system, the committee work that is the basis of this report would not have been possible.

William F. Eddy, *Chair*
Committee on National Statistics, 2009

Contents

Preface

The Committee on National Statistics (CNSTAT) has, since 1992, produced a report on principles and practices for a federal statistical agency, which draws from CNSTAT's many reports on specific agencies, programs, and topics. This report has been widely cited and used by Congress and federal agencies; it has helped shape legislation and executive actions to establish and evaluate statistical agencies; the U.S. Office of Management and Budget has cited it in regulatory language; and the U.S. Government Accountability Office has used it as a benchmark in reports to Congress. Statistical agencies have used it to inform new appointees, advisory committees, and others about what constitutes an effective and credible statistical organization. Beginning with the second edition in 2001, CNSTAT committed to updating the document every 4 years on a schedule to provide a current edition to newly appointed cabinet secretaries and other federal personnel at the beginning of each presidential administration (or second term).

This fourth edition presents and comments on four basic principles that statistical agencies must embody in order to carry out their mission fully: (1) They must produce data that are relevant to policy issues, (2) they must achieve and maintain credibility among data users, (3) they must achieve and maintain trust among data providers, and (4) they must achieve and maintain a strong position of independence from the appearance and reality of political control. The paper also discusses 11 important practices that are means for statistical agencies to live up to the four principles. These practices include a commitment to quality and professional practice and an active program of methodological and substantive research. The first three principles and 10 of the 11 practices have appeared in each prior edition; this fourth edition adds the principle that statistical agencies must operate from a strong position of independence and the practice that agencies must have ongoing internal and external evaluations of their programs.

The fourth edition retains the basic structure of previous editions in that Part I presents the principles and practices in summary form, and Part II, Commentary, further explains, defines, and illustrates the topics in Part I. The fourth edition includes new appendix material to orient the reader: Appendix A provides an overview of the organization of the U.S. federal statistical system, which is the most decentralized in the developed world, and compares the size of the system to the size of the federal government as a whole. Appendix B summarizes key legislation and regulations that affect federal statistical agencies, such as the Confidential Information Protection and Statistical Efficiency Act of 2002. Appendix C reproduces the Fundamental Principles of Official Statistics of the United Nations Statistical Commission. Appendix D provides addresses of Internet sites for major federal agencies that provide statistical data, and Appendix E reproduces the prefaces to the first, second, and third editions of the CNSTAT report.

We are sometimes asked what distinguishes a "principle" from a "practice." Although the distinction is not hard and fast, we deem "principles" to be fundamental and intrinsic to the concept of a federal statistical agency. Without policy relevance, credibility with data users, trust of data providers, and a strong position of independence, an agency cannot provide the benefits to policy makers and the public in a democratic society that are the rationale for establishing a *statistical* agency. We deem "practices" to be ways and means of making the basic principles operational and facilitating an agency's adherence to the basic principles.

Although focused on federal statistical agencies, many of the principles and practices articulated here likely also apply to statistical activities elsewhere, such as in federal policy, evaluation, research, and program agencies, in state and local government agencies, and in other countries. Finally, the principles and practices in this report remain guidelines, not prescriptions. We intend them to assist statistical agencies and to inform policy makers, data users, and others about the characteristics of statistical agencies that enable them to serve the public good.

William F. Eddy, *Chair*
Committee on National Statistics, 2009

Part I: Principles and Practices for a Federal Statistical Agency

Definition of a Federal Statistical Agency

Establishment of a Federal Statistical Agency

Principles for a Federal Statistical Agency
- Relevance to Policy Issues
- Credibility Among Data Users
- Trust Among Data Providers
- A Strong Position of Independence

Practices for a Federal Statistical Agency
- A Clearly Defined and Well-Accepted Mission
- Continual Development of More Useful Data
- Openness About Sources and Limitations of the Data Provided
- Wide Dissemination of Data
- Cooperation with Data Users
- Fair Treatment of Data Providers
- Commitment to Quality and Professional Standards of Practice
- An Active Research Program
- Professional Advancement of Staff
- A Strong Internal and External Evaluation Program
- Coordination and Cooperation with Other Statistical Agencies

NOTE: Part I is a summary statement of principles and practices for an effective statistical agency. Part II, Commentary, further explains, defines, and illustrates the topics in Part I.

DEFINITION OF A FEDERAL STATISTICAL AGENCY

A federal statistical agency is a unit of the federal government whose principal function is the compilation and analysis of data and the dissemination of information for statistical purposes.[1]

The theory and methods of the discipline of statistics and related fields and the practice of the profession of statistics are brought to bear on the compilation of data, on producing information from the data, and on disseminating that information.

- The *unit* is generally recognized as a distinct entity. It may be located within either a cabinet-level department or an independent agency, or it could itself be an independent agency.
- *Compilation* may include direct collection of data from individuals, organizations, or establishments or the acquisition of information from administrative records. It may also include assembling information from a variety of sources, including other statistical agencies, in order to produce an integrated data series, such as the national income and product accounts.
- *Analysis* may take various forms. It includes methodological research to improve the quality and usefulness of data. It also includes substantive analysis—for example, developing indicators, modeling, making projections, interpreting data, and explaining relationships among survey statistics at various levels of aggregation and other variables. Analysis by a statistical agency does not advocate policies or take partisan positions.
- *Dissemination* means making information available to the public, to the executive branch, and to Congress.
- *Statistical purposes* include description, evaluation, analysis, inference, and research. For these purposes, a statistical agency may collect data from individuals, establishments, or other organizations directly, or it may obtain data from administrative records, but it does not use these data for administrative, regulatory, or law enforcement purposes. Statistical purposes relate to descriptions of groups and exclude any interest in or identification of an individual person or economic unit. The data are used solely to

[1]The U.S. Office of Management and Budget (2007:33364) provides a similar definition of a statistical agency: "an agency or organizational unit of the executive branch whose activities are predominantly the collection, compilation, processing, or analysis of information for statistical purposes."

describe and analyze statistical patterns, trends, and relationships involving groups of persons or other units.

ESTABLISHMENT OF A FEDERAL STATISTICAL AGENCY

Statistics that are publicly available from government agencies are essential for a nation to advance the economic well-being and quality of life of its people. Public policy makers are best served by statistics that are relevant for policy decisions, accurate, timely, and credible. Individuals, households, corporations, academic institutions, and other organizations rely on high-quality, publicly available data as the basis for informed decisions on a wide variety of issues. Even more, the operation of a democratic system of government depends on the unhindered flow of statistical information that citizens can use to assess government actions and for other purposes. Federal statistical agencies are established to be a credible source of relevant, accurate, and timely statistics in one or more subject areas that are available to the public and policy makers.

"Relevant statistics" are statistics that measure things that matter to policy making and public understanding. Relevance requires concern for providing data that help users meet their current needs for decision making and analysis, as well as anticipating future needs. "Accurate statistics" are statistics that match the phenomena being measured and do so in repeated measurements. Accuracy requires proper concern for consistency across geographic areas and across time, as well as for statistical measures of errors in the data. "Timely statistics" are those that are known close in time to the phenomena they measure. Timeliness requires concern for issuing data as frequently as is needed to reflect important changes in what is being studied, as well as disseminating data as soon as practicable after they are collected. "Credibility" requires concern for both the reality and appearance of impartiality and independence from political control. It is the primary mission of agencies in the federal statistical system to work to ensure the relevance, accuracy, timeliness, and credibility of statistical information.

There are four major reasons to establish a statistical agency:

(1) The opportunity to achieve higher data quality and greater efficiency of statistical production through a consolidated and more highly professional activity.

(2) The need for ongoing, up-to-date information in a subject area

that extends beyond the scope of individual operating units, possibly involving other departments or agencies.

(3) The need to protect the confidentiality of responses of data providers, both individuals and organizations.

(4) The need for data series that are independent—not subject to control by policy makers or regulatory or enforcement agencies and equally available to all users.

The principles and practices for a federal statistical agency that are reviewed in this report pertain to individual agencies as separate organizational entities in the context of a decentralized system for providing federal statistics. Historically, the response of the U.S. government to new information needs has been to create separate statistical units, so that the United States now has one of the most decentralized statistical systems of any modern nation. This report does not comment on the advantages or disadvantages of the U.S. system nor compare it with other models for organizing government statistics. It discusses the need for federal statistical agencies to coordinate and cooperate with other agencies on a range of activities, describes the coordinating role of the U.S. Office of Management and Budget (OMB), and reviews some mechanisms for interagency collaboration. Appendixes A and B provide detailed information on the U.S. statistical system.

PRINCIPLES FOR A FEDERAL STATISTICAL AGENCY

Principle 1: Relevance to Policy Issues

A federal statistical agency must be in a position to provide objective information that is relevant to issues of public policy.

A statistical agency must be knowledgeable about the issues and requirements of public policy and federal programs and able to provide objective information that is relevant to policy and program needs. Objective information is information that is as accurate and comprehensible as possible and is not designed to promote a particular policy position or group interest. In establishing priorities for statistical programs for this purpose, a statistical agency must work closely with the users of such information in the executive branch, the Congress, and interested nongovernmental groups.

Statistical agencies must also provide objective information on the subject area(s) in their purview that is useful to a broad range of private- and public-sector users as well as the public. To establish priorities for such information, a statistical agency must maintain contact with a broad spectrum of users in the business sector, academia, state and local governments, and elsewhere.

Principle 2: Credibility Among Data Users

A federal statistical agency must have credibility with those who use its data and information.

It is essential that a statistical agency strive to maintain credibility for itself and for its data. Few data users are in a position to verify the completeness and accuracy of statistical information; they must rely on an agency's reputation as a credible source of accurate and useful statistics.

Credibility derives from the respect and trust of users for the statistical agency and its data. These qualities come from more than producing data that merit acceptance as objective, accurate, and timely. Respect and trust are earned when an agency exhibits openness about sources and limitations of the data provided, a willingness to understand and strive to meet user needs, even those users may not well articulate, and a posture of respect and trust in the users of its data.

To have credibility, an agency must be free—and must be perceived to be free—of political interference and policy advocacy. Also important for credibility is for an agency to follow such practices as wide dissemination of data on an equal basis to all users, openness about the data provided, and commitment to quality and professional practice, as well as a strong internal and external evaluation program to assess and improve its data systems.

Principle 3: Trust Among Data Providers

A federal statistical agency must have the trust of those whose information it obtains.

Data providers, such as respondents to surveys and custodians of administrative records, must be able to rely on the word of a statistical agency that the information they provide about themselves or others will be used only for statistical purposes. An agency earns the trust of its data providers by appropriately protecting the confidentiality of responses. Such protection, in particular, precludes the use of individually identifiable information maintained by a statistical agency—whether derived from survey responses or another agency's administrative records—for any administrative, regulatory, or law enforcement purpose.

Trust of data providers is also achieved by respecting their privacy. Such respect requires that an agency minimize the time and effort of people who provide information and inform them of the intended uses of the information. Trust of data providers is also achieved by successfully conveying to them the relevance of the data being collected for important public purposes. Data providers must be convinced not only that the information they provide will be kept confidential, but also that the information is intended for effective, beneficial public use.

Principle 4: A Strong Position of Independence

A federal statistical agency must have a strong position of independence within the government.

To be credible and unhindered in its mission to provide objective, useful, high-quality information, a statistical agency must not only be distinct from those parts of a department that carry out law enforcement and policy-making activities, but also have a widely acknowledged position of independence. It must be able to execute its mission without being subject to pressures to advance a political agenda. It must be impartial and avoid even the appearance that its collection, analysis, and reporting processes might be manipulated for political purposes or that individually identifiable data might be turned over for administrative, regulatory, or law enforcement purposes.

A strong degree of independence is reflected in such practices as adherence to predetermined schedules in the public release of important statistical indicators; control over statistical press releases; control over processing of the data that an agency collects; authority for professional decisions over the scope, content, and frequency of data compiled, analyzed, and disseminated; and maintaining a clear distinction between statistical information and policy interpretation. Without the credibility that comes from a strong degree of independence, users may lose confidence in the accuracy and objectivity of a statistical agency's data, and data providers may become less willing to cooperate with agency requests.

PRACTICES FOR A FEDERAL STATISTICAL AGENCY

The effective operation of a federal statistical agency must begin with a clearly defined and well-accepted mission. With this prerequisite, effective operation involves a wide range of practices: continual development of more useful data, openness about sources and limitations of the data provided, wide dissemination of data, cooperation with data users, fair treatment of data providers, commitment to quality and professional standards of practice, an active research program, professional advancement of staff, a strong internal and external evaluation program, and coordination and cooperation with other statistical agencies.

Practice 1: A Clearly Defined and Well-Accepted Mission

An agency's mission should include responsibility for all elements of its programs for providing statistical information—determining sources of data, measurement methods, efficient methods of data collection and processing, and appropriate methods of analysis—and ensuring the public availability not only of the data, but also of documentation of the methods used to obtain the data and their quality. The mission should include the responsibility for assessing information needs and priorities and ways to meet those needs, such as by the establishment, modification, or discontinuance of a survey, census, or other method of data collection, such as extracting information from administrative records.

Practice 2: Continual Development of More Useful Data

Statistical agencies should continually look to improve their data systems to provide information that is accurate, timely, and relevant for changing public policy needs. They should also continually seek to improve the efficiency of their programs for collecting, analyzing, and disseminating statistical information.

Ways for an agency to achieve these goals include the following:

- Seeking opportunities to combine data from multiple surveys or to integrate data from surveys with data from administrative records, with appropriate safeguards for confidentiality. When separate data sets are collected and analyzed in such a manner that they may be used together, the value of the resulting information and the efficiency of obtaining it may be greatly enhanced.

- Sharing technical information and ideas with other statistical agencies. Such sharing can stimulate the development of innovative data collection, analysis, and dissemination methods that improve the accuracy and timeliness of information and the efficiency of data operations.
- Establishing a multifaceted data collection program to provide relevant information for different types of data needs. Such a program could include one-time surveys on special topics, repeated surveys of cross-sections of the population that provide regularly updated statistics, and longitudinal surveys that track people, firms, and institutions over time and make it possible to analyze the causes and effects of changes in their circumstances.
- Using administrative records as part of an agency's data collection program.

Practice 3: Openness About Sources and Limitations of the Data Provided

A statistical agency should be open about its data and their strengths and limitations, taking as much care to understand and explain how its statistics may fall short of accuracy as it does to produce accurate data in the first place. Data releases from a statistical program should be accompanied by a full description of the purpose of the program; the methods and assumptions used for data collection, processing, and reporting; what is known and not known about the quality and relevance of the data; sufficient information for estimating variability in the data; appropriate methods for analysis that take account of variability and other sources of error; and the results of research on the methods and data.

When problems are found in a previously released statistic that could affect its use, an agency should issue a correction promptly and publicly. An agency should be proactive in seeking ways to alert known and likely users of the data about the nature of the problem and the appropriate corrective action.

Practice 4: Wide Dissemination of Data

A statistical agency should strive for the widest possible dissemination of the data it compiles. Data dissemination should be timely and public.

Also, measures should be taken to ensure that data are preserved and accessible for use in future years.

Elements of an effective dissemination program include the following:

- An established publications policy that describes, for a data collection program, the types of reports and other data releases to be made available, the audience to be served, and the frequency of release.
- A variety of avenues for data dissemination, chosen to reach as broad a public as reasonably possible. Channels of dissemination include, but are not limited to, an agency's Internet web site, government depository libraries, conference exhibits and programs, newsletters and journals, e-mail address lists, and the media for regular communication of major findings.
- Release of data in a variety of formats, including printed reports, easily accessible web site displays and databases, public-use microdata and other publicly available computer-readable files, so that the information can be accessed by users with varying skills and needs for data retrieval and analysis. All data releases should be suitably processed to protect confidentiality, with careful and complete documentation.
- For research and other statistical purposes, access to relevant information that is not publicly available through restricted access modes that protect confidentiality. Such modes include protected research data centers, remote monitored online access for special tabulations and analyses, and licensing of individual researchers to allow them to use confidential data on their desktop computers under stringent arrangements to ensure that no one else can access the information.
- Procedures for release of information that preclude actual or perceived political interference. In particular, the content and timing of the public release of data should be the responsibility of the statistical agency, and the agency or unit that produces the data should publish in advance and meet release schedules for important indicators to prevent even the appearance of manipulation of release dates for political purposes.
- Policies for the preservation of data that guide what data to retain and how they are to be archived for future secondary analysis.

Practice 5: Cooperation with Data Users

A statistical agency should consult with a broad spectrum of users of its data in order to make its products more useful. It should

- seek advice on data concepts, statistical methods, and data products from data users as well as from other professional and technical subject-matter and methodological experts, using a variety of formal and informal means of communication that are appropriate to the types of input sought;
- seek advice on its statistical programs and priorities from external groups, including those with relevant subject-matter and technical expertise;
- provide wide access to data while maintaining appropriate safeguards for the confidentiality of individual responses; and
- provide equal access to data to all users.

Practice 6: Fair Treatment of Data Providers

To maintain a relationship of respect and trust with data subjects and other data providers, a statistical agency should observe fair information practices. Such practices include the following:

- Policies and procedures to maintain the confidentiality of data, whether collected directly or obtained from administrative record sources, and to inform data providers of the manner and level of protection.
- Policies and procedures to inform data providers of the purposes of data collection and the anticipated uses of the information, whether their participation is mandatory or voluntary, and, if voluntary, using appropriate informed consent procedures to obtain their information.
- Respecting the privacy of respondents by minimizing the contribution of time and effort asked of them, consistent with the purposes of the data collection activity.
- Recognizing the value of respondents' participation in data collection programs by accurately representing the statistical information they provide and by making it widely and equally available to all.
- Seeking, to the extent practicable, input from respondents, as well as others, on useful information to collect and disseminate and on the best ways to obtain information.

Practice 7: Commitment to Quality and Professional Standards of Practice

A statistical agency should

- use modern statistical theory and sound statistical practice in all technical work;
- develop strong staff expertise in the disciplines relevant to its mission, in the theory and practice of statistics, and in data collection, processing, analysis, and dissemination techniques;
- develop an understanding of the validity and accuracy of its data and convey the resulting measures of quality to users in ways that are comprehensible to nonexperts;
- maintain quality assurance programs to improve data quality and to improve the processes of compiling, editing, and analyzing data;
- develop a strong and continuous relationship with appropriate professional organizations in the fields of statistics and relevant subject-matter areas; and
- document concepts, definitions, data collection methodologies, and measures of uncertainty and discuss possible sources of error in reports and other data releases.

Practice 8: An Active Research Program

A statistical agency should have a research program that is integral to its activities. Because smaller agencies may not be able to afford as extensive a research program as larger agencies, agencies should share research results and methods. Agencies can also augment their staff resources for research by obtaining the services of experts not on the agency's staff through consulting or other arrangements as appropriate.

The research program of a statistical agency should include the following:

- Research on the substantive issues for which the data were compiled. Such research should be conducted not only to provide useful objective analytical results, but also as a means to identify potential improvements to the content of the data, suggest improvements in the design and operation of the data collection, and provide fuller understanding of the limitations of the data.

- Research to evaluate and improve statistical methods, in particular the identification and creation of new statistical measures and the development of improved methods for analyzing errors in data that are due not only to sampling variability, but also to other sources. Research should also be conducted on ways to reduce the time and effort requested of respondents and to improve the timeliness, accuracy, and efficiency of data collection, analysis, and dissemination procedures.
- Research to understand how the agency's information is used, in order to make the data more relevant to policy concerns and more useful for policy research and decision making.

Practice 9: Professional Advancement of Staff

A statistical agency should recruit, develop, and support professional staff who are committed to the highest standards of quality work and professional practice. An agency's staff should also be committed to the highest standards of professional ethics with regard to maintaining the agency's credibility as an objective, independent source of accurate and useful information obtained through fair information practices.

To develop and maintain a high-caliber staff, a statistical agency must recruit qualified people with the relevant skills for its efficient and effective operation, including analysts in fields relevant to its mission (e.g., demographers, economists), statistical methodologists who specialize in data collection and analysis, and other specialized staff (e.g., computer specialists). To retain and make the most effective use of its staff, an agency should provide opportunities for work on challenging projects in addition to more routine, production-oriented assignments. An agency's personnel policies, supported with significant resources, should enable staff to extend their technical capabilities through appropriate professional and developmental activities, such as attendance and participation in professional meetings, participation in relevant training programs, and rotation of assignments. An agency should also seek opportunities to reinforce the commitment of its staff to ethical standards of practice.

Practice 10: A Strong Internal and External Evaluation Program

Statistical agencies should have regular, ongoing programs of evaluation for major statistical programs and program components and for the agency's portfolio of programs as a whole. Regular formal reviews of major data collection programs and their components should consider, among other topics, how to produce the highest quality data possible given the available resources. Regular formal reviews of an agency's portfolio should consider how to produce the most relevant information possible for policy makers and the public. Such evaluations should include internal reviews by staff and external reviews by independent bodies.

Practice 11: Coordination and Cooperation with Other Statistical Agencies

A statistical agency should seek opportunities to cooperate with other statistical agencies to enhance the value of its own information and that of other agencies in the federal statistical system. Although agencies differ in their subject-matter focus, there is overlap in their missions and a common interest in serving the public need for credible, high-quality statistics gathered as efficiently and fairly as possible.

When possible and appropriate, federal statistical agencies should cooperate not only with each other, but also with state and local statistical agencies in the provision of data for subnational areas. Federal statistical agencies should also cooperate with foreign and international statistical agencies to exchange information on both data and methods and to develop appropriate common classifications and procedures to promote international comparability of information.

Such cooperative activities as integrating data compiled by different statistical agencies invariably require effort to overcome differences in agency missions and operations. But the rewards are data more relevant to policy concerns and a stronger statistical system as a whole. For these reasons, statistical agencies must act as partners to one another, not only in the development of data, but also for the entire panoply of statistical activities, including definitions, concepts, measurement methods, analytical tools, professional practice, dissemination modes, means to protect the confidentiality of responses, and ways to advance the effective use of statistical information.

Part II: Commentary

This section comments on most of the topics in the principles and practices; the comments are offered to explain, illustrate, or further define the statement of principle in Part I.

DEFINITION OF A FEDERAL STATISTICAL AGENCY

A federal statistical agency is a unit of the federal government whose principal function is the compilation and analysis of data and the dissemination of information for statistical purposes.

A statistical agency may be labeled a bureau, center, division, or office or similar title, so long as it is recognized as a distinct entity. Statistical agencies have been established for several reasons: (1) to develop new information for an area of public concern (e.g., the Bureau of Labor Statistics, the National Center for Health Statistics); (2) to conduct large statistical collection and dissemination operations specified by law (e.g., the U.S. Census Bureau); (3) to compile and analyze statistics from sets of administrative records for policy purposes and public use (e.g., the Statistics of Income Division in the Internal Revenue Service); and (4) to develop broad and consistent estimates from a variety of statistical and administrative sources in accordance with a prespecified conceptual framework (e.g., the Bureau of Economic Analysis in the U.S. Department of Commerce). Once established, many statistical agencies engage in all of these functions to varying degrees.

This definition of a federal statistical agency does not include many statistical activities of the federal government because they are not performed by distinct units, or because they do not result in the dissemination of statistics to others—for example, statistics compiled by the U.S. Postal Service to set rates or by the U.S. Department of Defense to test weapons (see National Research Council, 1998b, 2002b, 2003b, 2006d, on statistics and testing for defense acquisition). Nor does it include agencies whose primary functions are the conduct or support of problem-oriented research, although their research may be based on information gathered by statistical means, and they may also sponsor important surveys, as do, for example, the National Institutes of Health, the Agency for Healthcare Research and Quality, and other agencies in the U.S. Department of Health and Human Services.

Finally, this definition of a statistical agency does not usually include agencies whose primary function is policy analysis and planning (e.g., the Office of Tax Analysis in the U.S. Department of the Treasury, the Office of the Assistant Secretary for Planning and Evaluation in the U.S. Department of Health and Human Services). Such agencies may collect and analyze statistical information, and statistical agencies, in turn, may perform some policy-related analysis (e.g., produce reports on trends in after-tax income or child care arrangements of families). However, to maintain credibility as an objective source of accurate, useful information, statistical agencies must be separate from units that are involved in developing policy and assessing policy alternatives.

The work of federal statistical agencies is coordinated through the Interagency Council on Statistical Policy (ICSP), created by the U.S. Office of Management and Budget (OMB) in the 1980s and authorized in statute in the 1995 reauthorization of the Paperwork Reduction Act. The ICSP is chaired by OMB and currently includes representation from a total of fourteen agencies and units, which are housed in nine cabinet departments and three independent agencies (see Appendix A):

- Bureau of Economic Analysis (Commerce Department)
- Bureau of Justice Statistics (Justice Department)
- Bureau of Labor Statistics (Labor Department)
- Bureau of Transportation Statistics (Transportation Department)
- Census Bureau (Commerce Department)
- Economic Research Service (Agriculture Department)
- Energy Information Administration (Energy Department)

- National Agricultural Statistics Service (Agriculture Department)
- National Center for Education Statistics (Education Department)
- National Center for Health Statistics (Health and Human Services Department)
- Office of Environmental Information (Environmental Protection Agency)
- Office of Research, Evaluation, and Statistics (Social Security Administration)
- Science Resources Statistics Division (National Science Foundation)
- Statistics of Income Division (Treasury Department)

Throughout the federal government, OMB recognizes more than 80 units and agencies that are not statistical agencies but that have annual budgets of $500,000 or more for statistical activities (U.S. Office of Management and Budget, 2008c:Table 1). The principles for federal statistical agencies presented here should apply to other federal agencies that carry out statistical activities, and they may find many of the detailed practices pertinent as well. Similarly, the principles and practices may be relevant to statistical units in state and local government agencies, and international audiences may also find them useful.

ESTABLISHMENT OF A FEDERAL STATISTICAL AGENCY

One of the most important reasons for establishing a statistical agency is to provide information that will allow for an informed citizenry. A democracy depends on an informed electorate. A citizen has a right to information that comes from a trustworthy, credible source and is relevant, accurate, and timely. Timely information of high quality is also critical to policy analysts and decision makers in both the public and private sectors. (For more information on the purposes of official statistics, see the Fundamental Principles of Official Statistics of the United Nations Statistical Commission in Appendix C; see also U.N. Economic Commission for Europe, 2003; U.N. Statistical Commission, 2003.) Federal statistical agencies serve the key functions of providing a broad array of information to the public and policy makers and of ensuring the necessary quality and credibility of the data.

Commercial, nonprofit, and academic organizations in the private sector also provide useful statistical information, including data they collect themselves and data they acquire from government agencies and other data collectors to which they add value. However, because the benefits of statistical information are shared widely throughout society and because it is often difficult to collect payments for these benefits, private markets are not likely to provide all of the data that are needed for public and private decision making or to make data as widely available as needed for important public purposes. Government statistical agencies are established to ensure that a broad range of information is publicly available. (See National Research Council, 1999b and 2005b, for a discussion of the governmental role in providing public goods, or near public goods, such as research and data.)

The United States government collected and published statistics long before any distinct federal statistical agency was formed (see Duncan and Shelton, 1978; Norwood, 1995). The U.S. Constitution mandated the conduct of a decennial census of population beginning in 1790, and the census enumeration was originally conducted by U.S. marshals as one of their many duties. Legislation providing for the compilation of statistics on agriculture, education, and income was enacted by Congress in the 1860s. The Bureau of Labor (forerunner of the Bureau of Labor Statistics) was established by law in 1884 as a separate agency with a general mandate to respond to widespread public demand for information on the conditions of industrial workers. The Census Bureau was established as a permanent agency in 1902 to conduct the decennial census and related statistical activities.

Many federal statistical agencies that can trace their roots back to the 19th or early 20th century, such as the National Center for Education Statistics and the National Center for Health Statistics, were organized in their current form following World War II. Several relatively new agencies have since been established, including the Energy Information Administration, the Bureau of Justice Statistics, and the Bureau of Transportation Statistics. In every case, the agency itself, in consultation with users of its information, has major responsibility for determining its specific statistical programs and for setting priorities. Initially, many of these agencies also had responsibilities for certain policy analysis functions for their department heads. More recently, policy analysis has generally been located in separate units that are not themselves considered to be statistical agencies, a separation that helps establish and maintain the credibility of statistical agencies as providers of data and analyses that are not designed for particular policy alternatives.

A statistical agency has at least two roles: (1) provider of the statistical information and analysis needed to inform policy making and program assessment by its own department, and (2) source of national statistics for the public in its area of concern. It is sometimes difficult to keep these two roles distinct on policy-relevant statistics. An effective statistical agency, nevertheless, will frequently play a creative, not just reactive, role in the development of data needed for policy analysis. Sometimes federal statistical agencies play additional roles, such as monitor and consultant on statistical matters to other units within the same department (see, e.g., National Research Council, 1985a) and collector of data on a reimbursable basis for other agencies.

There is no set rule or guideline for when it is appropriate to establish a separate federal statistical agency, carry on statistical activities within the operating units of departments and independent agencies, or contract for statistical services from existing federal statistical agencies or other organizations. Establishment of a federal statistical agency should be considered when one or more of the following conditions prevail:[1]

- There is a need for information on an ongoing basis beyond the capacity of existing operating units, possibly involving other departments and agencies. Such needs may require coordinating data from various sources, initiating new data collection programs to fill gaps, or developing regularly updated time series of estimates.
- There is a need, as a matter of credibility, to ensure that major data series are independent of policy makers' control.
- There is a need to establish the functional separation of data on individuals and organizations that are collected for statistical purposes from data on individuals and organizations that may be used for administrative, regulatory, or law enforcement uses. Such separation, recommended by the Privacy Protection Study Commission (1977), bolsters a culture and practice of respect for privacy and protection of confidentiality. Functional separation is easier to maintain when the data to be used for statistical purposes are compiled and controlled by a unit that is separate from operating units or department-wide data centers. The Confidential Information Protection and Statistical Efficiency Act of 2002 (CIPSEA) extended legal confidentiality protection to statistical data collections that may be carried

[1]The National Research Council (2001b:Ch. 6) cited a number of these reasons in recommending to the U.S. Department of Health and Human Services that it establish or identify a statistical unit to be assigned responsibility and authority for carrying out statistical functions and data collection for social welfare programs and the populations they serve; see also National Research Council and Institute of Medicine (2004).

out by any federal agency, whether a statistical agency or other type of agency (see Appendix B). Nonetheless, functional separation of statistical data from other kinds of data is important because it makes promises of confidentiality protection more credible.[2]

- There is a need to emphasize the principles and practices of an effective statistical agency, for example, professional practice, openness about the data provided, and wide dissemination of data.
- There is a need to encourage research and development of a broad range of statistics in a particular area of public interest or of government activity or responsibility.
- There is a need to consolidate compilation, analysis, and dissemination of statistics in one unit to encourage high-quality performance, eliminate duplication, and streamline operations.

PRINCIPLES FOR A FEDERAL STATISTICAL AGENCY

Principle 1: A federal statistical agency must be in a position to provide objective information that is relevant to issues of public policy.

A statistical agency supplies information not only for the use of managers and policy makers in the executive branch and for legislative designers and overseers in Congress, but also for all those who require objective statistical information on public issues, whether the information is needed for purposes of production, trade, consumption, or participation in civic affairs. Just as a free enterprise economic system depends on the availability of economic information to all participants, a democratic political system depends on—and has a fundamental duty to provide—wide access to information on education, health, transportation, the economy, the environment, criminal justice, and other social concerns.

Federal statistical agencies are responsible for providing statistics on conditions in a variety of areas. The resulting information is used both inside and outside the government not only to delineate problems and sometimes to suggest courses of action, but also to evaluate the results of

---

[2]Under the guidance issued for CIPSEA (see Appendix B), OMB in 2007 recognized two new statistical units: the Office of Applied Studies within the Substance Abuse and Mental Health Services Administration of the U.S. Department of Health and Human Services, and the Microeconomic Surveys Section of the Federal Reserve Board of Governors.

government activity or lack of activity. The statistics provide much of the basis on which the government itself is judged. This role places a heavy responsibility on federal statistical agencies for impartiality and objectivity.

In order to provide information that is relevant to public issues, statistical agencies need to reach out to users of the data. Federal statistical agencies usually are in touch with the primary users in their own departments. Considerable energy and initiative are required to open avenues of communication more broadly to other current and potential users, including analysts and policy makers in other federal departments, state and local government agencies, academic researchers, private-sector organizations, organized constituent groups, the media, and Congress. Advisory committees are recommended as a means to obtain the views of users, as well as people with relevant technical expertise (see, e.g., National Research Council, 1993a). Many agencies obtain advice from committees that are chartered under the Federal Advisory Committee Act—examples include the Advisory Committee on Agriculture Statistics for the National Agricultural Statistics Service, the Board of Scientific Counselors for the National Center for Health Statistics, and the Census Advisory Committee of Professional Associations for the Census Bureau. The Federal Economic Statistics Advisory Committee (FESAC), chartered in November 1999, provides substantive and technical advice to three agencies—the Bureau of Economic Analysis, the Bureau of Labor Statistics, and the Census Bureau—thereby providing an important cross-cutting perspective on major economic statistics programs (see http://www.bls.gov/bls/fesac.htm [December 2008]). Some agencies obtain advice from committees and working groups that are organized by an independent association, such as the American Statistical Association's Committee on Energy Statistics for the Energy Information Administration.

One frequently recommended method for alerting statistical agencies to emerging statistical information needs is for the agency's own staff to engage in analysis of its data (Martin, 1981; Norwood, 1995; Triplett, 1991). For example, relevant analysis may use the agency's data to examine correlates of key social or economic phenomena or to study the statistical error properties of the data. Such in-house analysis can lead to improvements in the quality of the statistics, to identification of new needs, to a reordering of priorities, and to closer cooperation and mutual understanding with policy analysis units. In its work for a policy analysis unit, a statistical agency describes conditions and possibly measures progress toward some previously identified goal, but it refrains from making policy recommendations. The distinction between statistical analysis and policy analysis is not always clear, and a statistical agency will need to consider carefully the extent of policy-related activities that are appropriate for it to undertake.

Principle 2: A federal statistical agency must have credibility with those who use its data and information.

Users of a statistical agency's data must be able to trust that the data were collected and analyzed in an objective, impartial manner and that they are as accurate and timely as the agency can make them. An agency should make every effort to provide accurate and credible statistics that will permit policy debates to be concerned about policy, not about the credibility of the data. Credibility is enhanced when an agency fully informs users of the strengths and weaknesses of the data, makes data available widely, and consults with users about priorities for data collection and analysis. When it does so, an agency is perceived to be working in the national interest, not the interest of a particular administration (Ryten, 1990).

Principle 3: A federal statistical agency must have the trust of those whose information it obtains.

The statistics programs of the federal government rely in large part on information supplied by individuals and by organizations outside the federal government, such as state and local governments, businesses, and other organizations. Some of this information is required by law or regulation (such as employers' wage reports), some of it is related to administration of government programs (such as information provided by benefit recipients), but much of it is obtained through the voluntary cooperation of respondents in statistical surveys. Even when response is mandatory, the cooperation of respondents reduces costs and likely promotes accuracy (see National Research Council, 1995b, 2004e). Important elements in encouraging such cooperation are that respondents believe that the data requested are important and legitimate for the government to collect, that they are being collected in an impartial, competent manner, and that the confidentiality of their responses will be protected.

In brief, trust in a statistical agency must be maintained, and the agency must not be perceived as being swayed by political considerations. Respondent trust also depends on providing respondents with realistic promises of confidentiality that the agency can reasonably expect to honor and then scrupulously honoring those promises. Finally, respondent trust depends on adopting practices that respect personal privacy, such as taking steps to minimize the intrusiveness of questions and the time and effort required to participate in a survey.

Principle 4: A federal statistical agency must have a strong position of independence within the government.

A statistical agency must be able to provide credible information that may be used to evaluate the program and policies of its own department or the government as a whole. More broadly, a statistical agency must be a trustworthy source of objective, accurate information for decision makers, analysts, and others inside and outside the government who want to use statistics to understand present conditions, draw comparisons with the past, and help guide plans for the future.[3] For these purposes, a strong position of independence for a statistical agency is essential.

Statistical agency independence must be exercised in a broad framework. Legislative authority usually gives ultimate responsibility to the secretary of the department rather than the statistical agency head. In addition, an agency is subject to the normal budgetary processes and to various coordinating and review functions of OMB, as well as the legislative mandates, oversight, and informal guidance of Congress.

Within this broad framework, a statistical agency must work to maintain its credibility as an impartial purveyor of information. In the long run, the effectiveness of an agency depends on its maintaining a reputation for impartiality; thus, an agency must be continually alert to possible infringements on its credibility and be prepared to argue strenuously against such infringements.

An agency head's independence can be strengthened by being appointed for a fixed term by the President, with approval by the Senate, as is the case with the heads of the Bureau of Labor Statistics and the National Center for Education Statistics. For a fixed term, it is desirable that it not coincide with the presidential term so that professional considerations are more likely to be paramount in the appointment process. In contrast, the heads of the Bureau of Justice Statistics, the Census Bureau, and the Energy Information Administration are presidential appointees, but their terms

[3]See the *Fundamental Principles of Official Statistics* of the United Nations Statistical Commission in Appendix C.

are not fixed and usually end with a change of administration. In some instances, heads of statistical agencies are career senior executives.

It is also desirable that a statistical agency head have direct access to the secretary of the department or the head of the independent agency in which the statistical agency is located. Such access allows the head to inform new secretaries about the appropriate role of a statistical agency and present the case for new statistical initiatives to the secretary directly. Among the agency heads with presidential appointments, such direct access currently is provided by legislation only for the Bureau of Labor Statistics and the Energy Information Administration.

It is desirable for a statistical agency to have its own funding appropriation from Congress and not be dependent on allocations from the budget of its parent department or agency, which may be subject to reallocation.

These organizational aspects—appointment of the agency head by the President with approval by the Senate for a fixed term not coincident with that of the administration, direct access to the secretary of the agency's department, and separate budgetary authority—are neither necessary nor sufficient for a strong position of independence for a statistical agency, but they facilitate such independence. In contrast, some agencies are under several layers of supervision within their departments (see Appendix A).

Other characteristics related to independence are that a statistical agency has the following:

- Authority for professional decisions over the scope, content, and frequency of data compiled, analyzed, or published within the framework set by its authorizing legislation. Most statistical agencies have such broad authority, limited by budgetary constraints, departmental requirements, OMB review, and congressional mandates.
- Authority for selection and promotion of professional, technical, and operational staff.
- Recognition by policy officials outside the statistical agency of its authority to release statistical information, including accompanying press releases and documentation, without prior clearance.
- Authority to control information technology systems for data processing and analysis in order to securely maintain the integrity and confidentiality of data and reliably support timely and accurate production of key statistics.
- Authority for the statistical agency head and qualified staff to speak about the agency's statistics before Congress, with congressional staff, and before public bodies.

- Adherence to fixed schedules in public release of important statistical indicators to prevent even the appearance of manipulation of release dates for political purposes.
- Maintenance of a clear distinction between statistical information and policy interpretations of such information by the president, the secretary of the department, or others in the executive branch.
- Dissemination policies that foster regular, frequent release of major findings from an agency's statistical programs to the public via the media, the Internet, and other means.

Control over personnel actions, especially the selection and appointment of qualified professional staff, including senior executive career staff, is an important aspect of independence. Agency staff reporting directly to the agency head should have formal education and deep experience in the substantive, methodological, operational, or management issues facing the agency as appropriate for their positions. In addition, professional qualifications are of the utmost importance for statistical agency heads, whether the profession is that of statistician or the subject-matter field of the statistical agency (National Research Council, 1997b). Relevant professional associations can be a source of valuable input on suitable candidates.

The authority to ensure that information technology systems fulfill the specialized needs of the statistical agency is another important aspect of independence. A statistical agency must be able to vouch for the integrity, confidentiality, and impartiality of the information collected and maintained under its authority so that it retains the trust of its data providers and data users. Such trust is fostered when a statistical agency has control over its information technology resources, and there is no opportunity or perception that policy, program, or regulatory agencies could gain access to records of individual respondents. A statistical agency also needs control over its information technology resources to support timely and accurate release of official statistics, which are often produced under stringent deadlines.

Authority to decide the scope and specific content of the data collected or compiled and to make decisions about technical aspects of data collection programs is yet another important element of independence, although such authority can never be without limits. Congress frequently specifies particular data that it wishes to be collected (e.g., data on job openings and labor turnover by the Bureau of Labor Statistics, data on family farms by the Economic Research Service and National Agricultural Statistics Service) and, in the case of the decennial census, requires an opportunity to review the proposed questions. The OMB Office of Information and Regulatory Affairs, under the Paperwork Reduction Act (and under the preceding Federal Reports Act), has the responsibility for designating a single data collection instrument for information wanted by two or more agencies. It also has the responsibility under the same act for reviewing all questionnaires and other instruments for the collection of data from 10 or more respondents (see Appendix B). In addition, the courts sometimes become involved in interpreting laws and regulations that affect statistical agencies, as in a number of issues concerning data confidentiality and Freedom of Information Act requests and in the use of sampling in the population census.

The budgetary constraints on statistical agencies and OMB review of data collections are ongoing; other pressures depend, in part at least, on the relations between a statistical agency and those who have supervisory or oversight functions. Agencies need to develop skills in communicating to oversight groups the need for statistical series and credibility in assessing the costs of statistical work. In turn, although it is standard practice for the secretary of a department or the head of an independent agency to have ultimate responsibility for all matters within the department or agency, the head of a statistical agency, for credibility, should be allowed full authority in professional and technical matters. For example, decisions to revise the methodology for calculating the consumer price index (CPI) or the gross domestic product (GDP) have been and are properly made by the relevant statistical agency heads.

Other aspects of independence that underscore a statistical agency's credibility are important as well. Authority to release statistical information and accompanying materials (including press releases) without prior clearance by department policy officials is important so that there is no opportunity for or perception of political manipulation of any of the information. Authority for the statistical agency head and qualified staff to speak about the agency's statistics before Congress, with congressional staff, and before public bodies is also important to bolster the agency's standing.

When a statistical agency releases information publicly, a clear distinction should be made between the statistical information and any policy interpretations of such. Not even the appearance of manipulation for political purposes should be allowed. This is one reason that statistical agencies are required by Statistical Policy Directive Number 3 (U.S. Office of Management and Budget, 1985) to adhere to predetermined schedules for the public release of key economic indicators and take steps to ensure that no person outside the agency can gain access to such indicators before the official release time. Statistical Policy Directive Number 4 (U.S. Office of Management and Budget, 2008b) requires agencies to develop and publish schedules for release of other important social and economic indicators as well (see Appendix B). When an agency modifies a customary release schedule for statistical purposes, it should announce and explain the change as far in advance as possible.

PRACTICES FOR A FEDERAL STATISTICAL AGENCY

Practice 1: A Clearly Defined and Well-Accepted Mission

A clear understanding of the mission of an agency, the scope of its statistical programs, and its authority and responsibilities are basic to planning and evaluating its programs and to maintaining credibility and independence from political control (National Research Council, 1986, 1997b). Some agency missions are clearly spelled out in legislation; other agencies have only very general legislative authority. On occasion, very specific requirements may be set by legislation or regulation.

Agencies should communicate their mission clearly to others. The use of the Internet is one means to publicize an agency's mission to a broad audience and to provide related information, including enabling legislation, the scope of the agency's statistical program, confidentiality provisions, operating procedures, and data quality guidelines. An agency's mission should focus on the compilation, evaluation, analysis, and dissemination of statistical information. In addition, considerable and formal attention must be paid to setting statistical priorities (National Research Council, 1976). Advice from outside groups should be sought on the agency's statistical program, on setting statistical priorities, on the statistical methods used, and on data products and services. Such advice may be sought in a variety of formal and informal ways, but it should be obtained from data users and providers as well as professional or technical experts in the subject-matter area and in statistical methods and procedures. A strong research program in the agency's subject-matter field can assist in setting priorities and identifying ways to improve an agency's statistical programs (Triplett, 1991).

Practice 2: Continual Development of More Useful Data

Federal statistical agencies cannot be static. To provide information of continued relevance for public and policy use, they must continually anticipate data needs for future policy considerations and look for ways to develop data systems that can serve broad purposes. To improve the quality and timeliness of their information, they must keep abreast of methodological and technological advances and be prepared to implement new procedures in a timely manner. They must also continually seek ways to make their operations more efficient. Preparing for the future requires that agencies reevaluate existing data series, plan new data series as required, and be innovative and open in their consideration of ways to improve their programs. Because of the decentralized nature of the federal statistical system, innovation often requires cross-agency collaboration. Innovation also implies a willingness to implement different kinds of data collection efforts to answer different needs.

Integration of Data Sources

One way to increase the usefulness of survey data is to integrate them with data from other surveys or with data from administrative records, such as social program records. Such integration typically requires that several agencies work together.

For example, in the area of health care provider statistics, a study by a panel of the Committee on National Statistics (CNSTAT) concluded that no single survey was likely ever to meet all the criteria, address all the technical problems, or meet all users' needs for data. In order to provide adequate information on the availability, financing, and quality of health care, a coordinated and integrated system of data collection activities involving several organizational entities was required (National Research Council and Institute of Medicine, 1992).

Similarly, a CNSTAT study on retirement income statistics concluded that some of the information that is essential for analysis of savings and retirement decisions and the effect of medical care use and expenditures on retirement income security is most efficiently and accurately obtained from existing administrative records (National Research Council, 1997a). To be useful for estimation, this information (e.g., Social Security earnings histories, Medicare and Medicaid benefits) must be linked to individual data that are available from such panel surveys as the Health and Retirement Study sponsored by the National Institute on Aging, the National Longitudinal Surveys sponsored by the Bureau of Labor Statistics, and the Census Bureau's Survey of Income and Program Participation. Similarly, linkage of employer and employment survey data with administrative records can provide enhanced analysis and modeling capability: a good example is the Census Bureau's Longitudinal Employer-Household Dynamics program (see http://lehd.did.census.gov/led [December 2008]; see also National Research Council, 2007a).

Challenges to cost-effective data collection from households and individuals because of declining survey response (see, e.g., de Leeuw and de Heer, 2002) make it more important than ever to consider ways in which administrative records can be used to bolster the completeness and quality of estimates from statistical agency programs while containing costs.[4] One or a combination of the following four approaches could be used: evaluate survey data against administrative data (taking cognizance of differences that could affect the comparisons and of sources of error in both sets of records); improve the methods used to impute values to survey nonrespondents on the basis of patterns in administrative data; substitute administrative data for survey data; and combine survey and administrative data in statistical models for specific estimates (see National Research Council, 2000c, 2000d). In most uses of administrative data, not only must consideration be given to upfront investments to facilitate the most effective approach to their use, but also careful attention must be paid to the means by which the confidentiality of linked or augmented data files can be protected while allowing access for research purposes (National Research Council, 2005b). Care must also be taken to ensure that extracts of data from administrative records were prepared correctly according to the specifications provided by the statistical agency.

Sharing of Microdata

Another way to improve data quality and develop new kinds of information is for statistical agencies that collect similar information to share microdata records. For example, the sharing of business data would make it possible to evaluate reporting errors and the completeness of coverage of business firms in different surveys. Such sharing would also make it possible to develop more useful and accurate statistics on the nation's economy while decreasing the reporting burden on business data providers (National

[4]Lower response rates reduce the effective sample size and increase the sampling error of estimates from surveys; lower rates also increase response bias in survey estimates to the extent that nonrespondents differ from respondents in ways that affect analysis and are not addressed by weighting and imputation procedures.

Research Council, 2006b). Subtitle B of CIPSEA, for the first time in the nation's history, authorizes the sharing of business data among the three principal statistical agencies that produce the nation's key economic statistics—the Bureau of Economic Analysis (BEA), the Bureau of Labor Statistics (BLS), and the U.S. Census Bureau.[5] The first formal proposal for data sharing under CIPSEA involved matching data from BEA's international investment surveys with data from the Census Bureau's Survey of Industrial Research and Development conducted for the National Science Foundation. The results helped BEA improve its survey sample frames and enabled the Census Bureau to identify companies that were not previously known to engage in research and development activities (U.S. Office of Management and Budget, 2004b:44-45).

Longitudinal Data

The need to understand temporal changes in important social or economic events may call for the development of longitudinal surveys that track people, institutions, or firms over time. Developing longitudinal data (and general purpose repeated cross-sectional data, as well) usually requires much coordination with policy research agencies, other statistical agencies, and academic researchers. Longitudinal data may require more sophisticated methods for collection and analysis than data from repeated or one-time cross-sectional surveys. In addition, considerable time may be needed to produce useful data products for analyzing transitions and other dynamic characteristics of longitudinal samples (although production of cross-sectional products from longitudinal surveys need not take long). Yet data from longitudinal surveys are potentially very useful—sometimes, they are the only means to answer important policy questions (see, e.g., National Research Council, 1997a, on data needs to inform retirement income policy, and National Research Council, 2001b, on data needs to evaluate the effects of the 1996 welfare reform legislation).

Historically, because statistical agencies are oriented toward the mission of their particular department, the longitudinal surveys they developed (and cross-sectional data activities as well) typically focused on subject matter and population groups (or other entities) that the department serves. For

[5]The Census Bureau cannot share with BEA or BLS any tax information of businesses or individuals that it has permission to acquire from the Internal Revenue Service for statistical purposes without revision of Title 26 of the U.S. Code.

example, separate data sets are available on health characteristics of infants and children, educational characteristics for children and teenagers, and work force characteristics for adults. Increasingly, however, agencies have considered surveys that follow individuals across such key transitions as from early childhood to school and from school to the labor force (National Research Council, 1998a; National Research Council and Institute of Medicine, 2004).

Examples of statistical agency surveys that are designed for analysis of some kinds of transitions include the Early Childhood Longitudinal Study (ECLS), sponsored by the National Center for Education Statistics in collaboration with other agencies, and the National Longitudinal Surveys of Youth (NLSY79, NLSY97), sponsored by the Bureau of Labor Statistics. The ECLS includes two cohorts of children, one of kindergartners in 1998 who were followed through eighth grade and another of babies born in 2001 who were followed through kindergarten (http://www.nces.edu.gov/ecls [December 2008]). A new cohort of kindergartners will be sampled in fall 2010 and followed through fifth grade. The NLSY includes two cohorts of young people, one of people ages 14-22 in 1979, who are being interviewed every other year, and the other of people ages 12-17 in 1997, who are being interviewed annually (http://www.bls.gov/nls/home.htm [December 2008]).

Other important longitudinal surveys are sponsored by research agencies—for example, the National Institute on Aging sponsors the Health and Retirement Study (HRS), and the National Institute of Child Health and Human Development sponsors the new National Children's Study (NCS) (see National Research Council and Institute of Medicine, 2008). The HRS, which began in 1992, includes people aged 50 and older, who are interviewed every 2 years, with a new cohort introduced every 6 years (http://hrsonline.isr.umich.edu [December 2008]). The NCS will include 100,000 children and follow them and their families from before birth through age 21, with enrollment at the first sites in January 2009 (http://www.nationalchildrensstudy.gov [December 2008]).

Operational Methods

It is important for statistical agencies to be innovative in the methods used for data collection, processing, estimation, analysis, and dissemination. Agencies need to investigate new or modified methods that have the potential to improve the accuracy and timeliness of their data and the efficiency of their operations. Careful evaluation of new methods is required to assess their benefits and costs in comparison with current methods and to determine effective implementation strategies, including the development of methods for bridging time series before and after a change in procedures.

For example, experience with the use of computer-assisted interviewing techniques, which many agencies have adopted for data collection, has identified benefits. It has also identified challenges for the timely provision of data and documentation that require continued research to develop solutions that maximize the gains from these techniques (see National Research Council, 2003e).

Statistical agencies have turned to the Internet as a standard vehicle for data dissemination and are increasingly using it as a means of data collection. Internet dissemination facilitates the timely availability of data to a broad audience and provides a valuable tool for users to learn of related data sets from other agencies. However, it poses challenges in several areas, such as how best to provide information on data quality and appropriate use of the data to an audience that spans a wide range of analytical skills and understanding.

Internet data collection poses new challenges in such areas as sample design, questionnaire design, and protecting data confidentiality. Yet even as work is ongoing on meeting these challenges, population censuses around the world, federal business surveys, and other surveys are using the Internet as one data collection mode to reduce costs and facilitate response (see National Research Council, 2008a, on Internet use in population censuses). The use of the Internet also requires careful evaluation of the effects on the quality of responses in comparison with traditional data collection modes (telephone, mail, personal interview).

Practice 3: Openness About Sources and Limitations of the Data Provided

A critically important means to instill credibility and trust among data users and data providers is for an agency to operate in an open and fully transparent manner with regard to the sources and the limitations of its data. Openness requires that an agency provide a detailed description of its data with acknowledgment of any uncertainty and a description of the methods used and assumptions made. Agencies should provide to users reliable indications of the kinds and amounts of statistical error to which the data are subject (see Brackstone, 1999; Federal Committee on Statistical

Methodology, 2001a; see also President's Commission on Federal Statistics, 1971). Some statistical agencies have developed detailed quality profiles for some of their major series, such as those developed for the American Housing Survey (Chakrabarty, 1996), the Residential Energy Consumption Survey (Energy Information Administration, 1996), the Schools and Staffing Survey (Kalton et al., 2000), and the Survey of Income and Program Participation (U.S. Census Bureau, 1998). Earlier, the Federal Committee on Statistical Methodology (1978c) developed a quality profile for employment as measured in the Current Population Survey. These profiles have proved helpful to experienced users and agency personnel responsible for the design and operation of major surveys and data series (see National Research Council, 1993a, 2007b).

Openness about data limitations requires much more than providing estimates of sampling error. In addition to a discussion of aspects that statisticians characterize as nonsampling errors, such as coverage errors, nonresponse, measurement errors, and processing errors, a description of the concepts used and how they relate to the major uses of the data is desirable. Descriptions of the shortcomings of and problems with the data should be provided in sufficient detail to permit the user to take them into account in analysis and interpretation. Descriptions of how the data relate to similar data collected by other agencies should also be provided, particularly when the estimates from two or more series differ significantly in ways that may have policy implications.

Openness means that a statistical agency should describe how decisions on methods and procedures were made for a data collection program. It is important to be open about research conducted on methods and data and other factors that were weighed in a decision.

Openness also means that, when mistakes are discovered after statistics are released, the agency has an obligation to issue corrections publicly and in a timely manner. The agency should use not only the same dissemination vehicles to announce corrections that it used to release the original statistics, but also use additional vehicles, as appropriate, to alert the widest possible audience of current and future users of the corrections in the information.

In summary, agencies should make an effort to provide information on the quality, limitations, and appropriate use of their data that is as frank and complete as possible. Such information, which is sometimes termed "metadata," should be made available in ways that are easy for users to access and understand, recognizing that users differ in their level of understanding of statistical data (see National Research Council, 1993a, 1997b,

2007b). Agencies need to work to educate users that all data contain some uncertainty and error, which does not mean the data are wrong but that they must be used with care.

The Information Quality Act of 2000 stimulated all federal agencies to develop written guidelines for maintaining and documenting the quality of their information programs and activities. Using a framework developed collaboratively by the members of the Interagency Council on Statistical Policy (U.S. Departments of Agriculture et al., 2002), individual statistical agencies have developed quality guidelines for their own data collection programs, which are available on the Internet (see Practice 7 and Appendix B).

Practice 4: Wide Dissemination of Data

A statistical agency must have vigorous and well-planned dissemination programs to get information into the hands of users who need it on a timely basis. Planning should be undertaken from the viewpoint that the public has contributed the data elements, has paid for the data collection and processing, and should in return have the information accessible in ways that make it as useful as possible to the largest number of users.

A good dissemination program provides data to users in forms that are suited to their needs. Data release may take the form of regularly updated time series, cross-tabulations of aggregate characteristics of respondents, and analytical reports that are made available in printed publications, on computer-readable media (e.g., CD-ROM), and on the Internet (see Appendix D).

Yet another form of dissemination involves access to microdata files, which make it possible to conduct in-depth research in ways that are not possible with aggregate data. Public-use microdata files may be developed for general release. Such files contain data for individual respondents that have been processed to protect confidentiality by deleting, aggregating, or modifying any information that might permit individual identification. Alternatively, an agency may provide or arrange for a facility on the Internet to allow users to aggregate individual microdata to suit their purposes, with safeguards so that the data cannot be retabulated in ways that could identify individual respondents. Another alternative is to grant a license to individual researchers to analyze restricted microdata (that is, data that have not been processed for general release) at their own sites by agreeing to follow strict procedures for protecting confidentiality and accepting liability for penalties if confidentiality is breached. A fourth alternative is to allow researchers to analyze restricted microdata at secure sites maintained by a statistical agency, such as one of the Census Bureau's Research Data Centers located at several universities and research organizations around the country or the National Center for Health Statistics' Research Data Center at its headquarters (see Doyle et al., 2001; National Research Council, 2005b). Agencies should consider all forms of dissemination in order to gain the most use of their data consistent with protecting the confidentiality of responses.

The stunning improvements over the past two decades in computing speed, power, and storage capacity, the growing availability of information from a wide range of public and private sources on the Internet, and the increasing richness of statistical agency data collections have increased the risk that individually identifiable information can be obtained (see National Research Council, 2003d:Ch. 5, 2005b). Statistical agencies must be vigilant in their efforts to protect against the increased threats to disclosure from their summary data and microdata products while honoring their obligation to be proactive in seeking ways to provide data to users. When statistical data are not disseminated in useful forms, there is a loss to the public, not only of wasted taxpayer dollars, but also of research findings that could have informed public policy and served other important societal purposes.

A good dissemination program for statistical data uses a variety of channels to inform the broadest possible audience of potential users about available data products and how to obtain them. Such channels may include providing direct access to data on the Internet, depositing data products in libraries, establishing a network of data centers (such as the Census Bureau's state data centers and the National Agricultural Statistics Service's field offices), holding exhibits and making presentations at conferences, and maintaining lists of individuals and organizations to notify of new data. Agencies should also arrange for archiving of data with the National Archives and Records Administration (NARA) and other data archives, as appropriate, so that data are available for historical research in future years with suitable protections for confidentiality.

An effective dissemination program provides not only the data, but also information about the strengths and weaknesses of the data in ways that can be comprehended by diverse audiences. Information about the limitations of the data should be included in every form of data release, whether in a printed report, on a computer-readable data file, or on the Internet.

On occasion, the objective of presenting the most accurate data possible may require more time than is consistent with the needs of users for the information. The tension between frequency and promptness of release on one hand and accuracy on the other should be explicitly considered. When concerns for timeliness prompt the release of preliminary estimates (as in some economic indicators), consideration should be given to the frequency of revisions and the mode of presentation of revised figures from the point of view of the users as well as the issuers of the data. Agencies that release preliminary estimates must educate the public about differences among preliminary, revised, and final estimates.

Practice 5: Cooperation with Data Users

Users of federal statistical data span a broad spectrum of interests and needs. They include policy makers, planners, administrators, and researchers in federal agencies, state and local governments, the business sector, and academia. They also include activists, citizens, students, and media representatives. An effective statistical agency endeavors to learn about its data users and to obtain input from them on the agency's statistical programs.

The needs of users can be explored by forming advisory committees, holding focus groups, analyzing requests and Internet activity, or undertaking formal surveys of users. The task requires continual alertness to the changing composition and needs of users and the existence of potential users. An agency should cooperate with professional associations, institutes, universities, and scholars in the relevant fields to determine the needs of the research community and obtain their insight on potential uses. An agency should also work with relevant associations and other organizations to determine the needs of business and industry for its data.

Within the limitations of its confidentiality procedures as noted above, an agency should seek to provide maximum access to its data, including making the data available to external researchers for secondary analysis (National Research Council, 1985c, 2005b). Having data accessible for a wide range of analyses increases the return on the investment in data collection and provides support for an agency's program. Once statistical data are made public, they may be used in numerous ways not originally envisaged. An agency should attempt to monitor the major uses of its data as part of its efforts to keep abreast of user needs. In 2002 OMB introduced requirements for performance assessment of federal agencies; for statistical agencies, the requirements emphasize assessment of how well the agency understands and serves its users (see Appendix B).

Researchers and other users of data frequently request data from statistical agencies for specific purposes. The agency should have procedures in place for referring users to professionals within the agency who can comprehend the user's purposes and needs and who have a thorough knowledge of the agency's data. Statistical agencies should view these services as a part of their dissemination activities.

Ensuring equal access requires avoiding release of data to selected individuals or organizations in advance of other users. Agencies that prepare special tabulations of their data on request for external groups must be alert to the proposed uses. If the data are to be used in court cases, administrative proceedings, or collective bargaining negotiations, it is wise to have a known policy ensuring that all sides may receive the special tabulations, regardless of which side requested them or paid the cost of the tabulation.

Practice 6: Fair Treatment of Data Providers

Clear policies and effective procedures for protecting data confidentiality, respecting the privacy of respondents, and, more broadly, protecting the rights of human research participants are critical to maintaining the quality and comprehensiveness of the data that federal statistical agencies provide to policy makers and the public. Part of the challenge for statistical agencies is to develop effective means of communicating not only the agency's protection procedures and policies, but also the importance of the data being collected for the public good.

Protecting Confidentiality

Data providers must believe that the data they give to a statistical agency will not be used by the agency to harm them. For statistical data collection programs, protecting the confidentiality of individual responses is considered essential to encourage high response rates and accuracy of response. (For reviews of research on the relationship of concerns about confidentiality protection to response rates, see Hillygus et al., 2006; National Research Council, 2004e:Ch. 4.) Furthermore, if participants have been assured of confidentiality, then under federal policy for the protection of human subjects, disclosure of identifiable information about them would violate the principle of respect for persons even if the information is not sensitive and would not result in any social, economic, legal, or other harm (National Research Council, 2003d:Ch. 5).

Historically, some agencies had legislative mandates supporting promises of confidentiality (e.g., for the U.S. Census Bureau, Title 13 of the U.S. Code, first enacted in 1929, and for the National Agricultural Statistics Service, various provisions in Title 7 of the U.S. Code); other agencies (e.g., the Bureau of Labor Statistics) relied on strong statements of policy, legal precedents in court cases, or custom (see Gates, 2000; Norwood, 1995). The latter agencies risked having their policies overturned by judicial interpretations of legislation or executive decisions that might have required the agency to disclose identifiable data collected under a pledge of confidentiality (for an example involving the Energy Information Administration, see National Research Council, 1993b:185-186).

To give additional weight and stature to policies that statistical agencies had pursued for decades, OMB issued a Federal Statistical Confidentiality Order on June 27, 1997. This order assured respondents who provided statistical information to specified agencies that their responses would be held in confidence and would not be used against them in any government action, "unless otherwise compelled by law" (U.S. Office of Management and Budget, 1997; see also Appendix B).

CIPSEA became law in 2002, as Title V of the E-Government Act of 2002. Subtitle A of CIPSEA provides a statutory basis for protecting the confidentiality of all federal data collected for statistical purposes under a confidentiality pledge, including but not limited to data collected by statistical agencies. Subtitle A places strict limits on the disclosure of individually identified information collected with a pledge of confidentiality; such disclosure to persons other than the employees or agents of the agency collecting the data can occur only with the informed consent of the respondent and the authorization of the agency head and only when the disclosure is not prohibited by any other law (e.g., Title 13). It also provides penalties for employees or agents who knowingly or willfully disclose statistical information (up to 5 years in prison, up to $250,000 in fines, or both). OMB issued guidance in 2007 to assist agencies in implementing Subtitle A of CIPSEA (U.S. Office of Management and Budget, 2007; see also Appendix B).

Although confidentiality protection for statistical data is now on a much firmer legal footing across the federal government than prior to CIPSEA, there is an exception for some data from the National Center for Education Statistics (NCES) that could have an adverse effect on survey response. The USA PATRIOT Act of 2001, Section 508, amended the National Center for Education Statistics Act of 1994 to allow the U.S. Attorney General (or an assistant attorney general) to apply to a court to obtain any "reports, records, and information (including individually identifiable information) in the possession" of NCES that are considered relevant to an authorized investigation or prosecution of domestic or international terrorism. Section 508 also removed the penalties for NCES employees who furnish individual records under this section.

Statistical agencies continually strive to avoid inadvertent disclosure of confidential information in disseminating data. Recently, the widespread dissemination of statistical data via the Internet has heightened attention by agencies to ensuring that effective safeguards to protect confidential information are in place. Risks are increased when data for small groups are tabulated, when the same data are tabulated in a variety of ways, or when public-use microdata files (samples of records for unidentified individuals or units) are released with highly detailed content. Longitudinal surveys, for example, particularly newer ones, typically have richly detailed content for multiple domains (e.g., health, education, labor force participation) or multiple respondents (e.g., parents, students, teachers) or both. Risks may also be increased when surveys include linked administrative data or collect biomarkers from blood samples or other physiological measures (National Research Council, 2001a).

Because of the disclosure risks associated with detailed tabulations and rich public-use microdata files, there is always a tension between the desire to safeguard confidentiality and the desire to provide public access to data. This dilemma is an important one to federal statistical agencies, and it has stimulated ongoing efforts to develop new statistical and administrative procedures to safeguard confidentiality while permitting more extensive access. An effective federal statistical agency will exercise judgment in determining which of these procedures are best suited to its requirements to serve data users while protecting confidentiality. (Several Committee on National Statistics study panels have discussed these issues and alternative procedures for providing data access while maintaining confidentiality protection; see National Research Council, 1993b, 2000a, 2003d, 2005b.)

Respecting Privacy

To promote trust and encourage accurate response from data providers, it is important that statistical agencies respect their privacy. When data providers are asked to participate in a survey, they should be told whether the survey is mandatory or voluntary, how the data will be used, and who will have access to the data. In the case of voluntary surveys, information on these matters is necessary in order for data providers to give their informed consent to participate (see National Research Council, 2003d, on regulations and procedures for informed consent).

Respondents invest time and effort in replying to surveys. The amount of effort or burden varies considerably from survey to survey, depending on such factors as the complexity of the information that is requested. Statistical agencies should attempt to minimize such effort, to the extent possible, by using concepts and definitions that fit respondents' common understanding; by simplifying questionnaires; by allowing alternative modes of response (e.g., via the Internet) when appropriate; and by using administrative records or other data sources, if they are sufficiently complete and accurate to provide some or all of the needed information. In surveys of businesses or other institutions, agencies should seek innovative ways to obtain information from the institution's records and minimize the need for respondents to reprocess and reclassify information. It is also the responsibility of agencies to use qualified, well-trained interviewers. Respondents should be informed of the likely duration of a survey interview and, if the survey involves more than one interview, how many times they will be contacted over the life of the survey. This information is particularly important when respondents are asked to cooperate in extensive interviews, search for records, or participate in longitudinal surveys.

Ways in which participation in surveys can be made easier for respondents and result in more accurate data can be explored by such means as focus group discussions or surveys. Many agencies apply the principles of cognitive psychology to questionnaire design, not only to make the resulting data more accurate, but also to make the time and effort of respondents more efficient (National Research Council, 1984). Some agencies thank respondents for their cooperation by providing them with brief summaries of the information after the survey is compiled.

Increasing privacy concerns may contribute to observed declines in survey response rates. In a time when individuals are inundated with requests for information from public and private sources, when there are documented instances of identity theft and other abuses of confidential information on the Internet, when individual information is being used for terrorism-related investigatory or law enforcement purposes, it may not be surprising that individuals object to responding to censuses and surveys, even when the questions appear noninvasive and the data are collected for statistical purposes under a pledge of confidentiality. (See National Research Council, 2008b, for a literature review of public opinion on privacy in the wake of the September 11, 2001, terrorist attacks [Appendix M], and for a conclusion [p. 84] that "census and survey data collected by the federal statistical agencies are not useful for terrorism prevention.")

The E-Government Act of 2002 requires agencies to develop privacy impact assessments (PIAs) whenever ". . . initiating a new collection of information . . . in an identifiable form. . . ." The purpose of a privacy impact assessment is to ensure there is no collection, storage, access, use, or dissemination of identifiable information that is not both needed and permitted. In response, statistical agencies have begun conducting and releasing PIAs for statistical programs and, in the process, rethinking how to respect individual privacy in order to maintain trust with data providers (see Appendix B).

Statistical agencies should devote resources to understanding the privacy and confidentiality concerns of individuals (and organizations). They should also devote resources to devising effective strategies for communicating privacy and confidentiality policies and practices to respondents. Such strategies appear to be more necessary—and more challenging—than ever before.

Finally, a reason that respondents reply to statistical surveys is that they believe that their answers will be useful to the government or to society generally. Statistical agencies should respect this contribution by compiling the data and making them accessible to users in convenient forms. A statistical agency has an obligation to publish statistical information from the data it has collected unless it finds the results invalid.

Protecting Human Research Participants

Collecting data from individuals as part of a research study or a statistical information program is a form of research involving human participants, for which the federal government has developed regulations, principles, and best practices over a period of 50 years (National Research Council, 2003d). The pertinent regulations, which have been adopted by 10 departments and 7 agencies, are known as the "Common Rule" (45 CFR §46). The Common Rule regulations require that researchers protect the privacy of human participants and maintain the confidentiality of data collected from them, minimize the risks to participants from the data collection and analysis, select participants equitably with regard to the benefits and risks of the research, and seek informed consent from participants. Under the regulations, most federally funded research involving human participants must be reviewed by an independent institutional review board (IRB) to determine that the design meets the ethical requirements for protection. (For information about the Common Rule and procedures for the certification of IRBs by the Office for Human Research Protections in the U.S. Department of Health and Human Services, see http://www.hhs.gov/ohrp [December 2008].)

Data collections of federal statistical agencies are subject to IRB review within some departments. The Census Bureau, citing the confidentiality provisions in its own enabling legislation (13 USC §9), has maintained an exemption from IRB review for its data collection programs under 45 CFR §46.101(b.3), which permits exemption if "federal statute(s) require(s) without exception that the confidentiality of the personally identifiable information will be maintained throughout the research and thereafter."

Whether or not a statistical agency is subject to formal IRB review, it should strive to incorporate the spirit of the Common Rule regulations in the design and operation of its data collection programs. An agency that is required to obtain IRB approval for data collection should work proactively with the IRB to determine how best to apply the regulations in ways that do not unnecessarily inhibit response. For example, signed written consent is not necessary for mail surveys and is hardly ever necessary for telephone surveys of the general population; such documentation does not provide any added protection to the respondent, and it is likely to reduce participation. As noted above, an effective statistical agency will seek ways—such as sending an advance letter—to furnish information to potential respondents that will help them make an informed decision about whether to participate. Such information should include the planned uses of the data and their benefits to individuals and the public.

Practice 7: Commitment to Quality and Professional Standards of Practice

The best guarantee of high-quality data is a strong professional staff that includes experts in the subject-matter fields covered by the agency's program, experts in statistical methods and techniques, and experts in data collection, processing, and other operations. A major function of an agency's leadership is to strike a balance among these groups and promote working relationships that make the agency's program as productive as possible, with each group of experts contributing to the work of the others.

An effective statistical agency devotes resources to developing, implementing, and inculcating standards for data quality and professional practice. Although a long-standing culture of data quality contributes to professional practice, an agency should also seek to develop and document standards through an explicit process. The existence of explicit standards and guidelines, regularly reviewed and updated, facilitates training of new in-house staff and contractors' staffs. The OMB document, *Standards and Guidelines for Statistical Surveys* (U.S. Office of Management and Budget, 2006b), is helpful in that it covers every aspect of a survey from planning through data release (see also U.S. Office of Management and Budget, 2006a, and Appendix B).[6] It recommends that agencies develop additional, more detailed standards that focus on their specific statistical activities (see, e.g., the Statistical Standards of the National Center for Education Statistics, available at http://nces.ed.gov/statprog/2002/stdtoc.asp [December 2008]; and the Energy Information Administration's Standards Manual, available at http://www.eia.doe.gov/smg/Standards.html [December 2008]).

An effective statistical agency keeps up to date on developments in theory and practice that may be relevant to its program, such as new techniques for imputing missing data (see, e.g., National Research Council, 2004e:App. F). An effective agency is also alert to changes in the economy or in society that may call for changes in the concepts or methods used in particular data sets.[7] Yet the need for change often conflicts with the need for comparability with past data series, and this issue can easily dominate consideration of proposals for change. Agencies have the responsibility to manage this conflict by initiating more relevant data series or revising existing series to improve quality while providing information to compare old and new series, such as was done when the BLS revised the treatment of owner-occupied housing in the CPI.

To ensure the quality of its data collection programs and reports, an effective statistical agency has mechanisms and processes for obtaining both inside and outside review of such aspects as the soundness of the data collec-

---

[6]The data quality guidelines of statistical agencies in other countries are also helpful; for Canada, see http://www.statcan.ca [December 2008]; for Great Britain, see http://statistics.gov.uk [December 2008].

[7]Reviews of concepts underlying important statistical data series include: National Research Council (1995a and 2005c) on concepts of poverty; National Research Council (2002a) on cost-of-living concepts; National Research Council (2005a) on "satellite" accounts for nonmarket activities, such as home production, volunteerism, and human capital investment; National Research Council (2006a) on concepts of food insecurity and hunger; and National Research Council (2006c) on concepts of residence for the U.S. census and the American Community Survey.

tion and estimation methods and the completeness of the documentation of the methods used and the error properties of the data. For individual publications and reports, formal processes are needed that incorporate review by agency technical experts and, as appropriate, by technical experts in other agencies and outside the government. (See Appendix B for a description of recent OMB guidelines for peer review of scientific information; reviews at a program or agency-wide level are considered under Practice 10.)

Practice 8: An Active Research Program

Substantive Research and Analysis

A statistical agency should include staff with responsibility for conducting objective substantive analyses of the data that the agency compiles, such as analyses that assess trends over time or compare population groups:

- Agency analysts are in a position to understand the need for and purposes of the data from a survey or other data collection program and know how the statistics will be used. Such information must be available to the agency and understood thoroughly if the survey design is to produce the data required.
- Those involved in analysis can best articulate the concepts that should form the basic framework of a statistical series. Agency analysts are well situated to understand and transmit the views of external users and researchers; at the same time, close working relationships between analysts and data producers are needed for the translation of the conceptual framework into the design and operation of the survey or other data collection program.
- Agency analysts have access to the complete microdata and so are in a better position than analysts outside the agency to understand and describe the limitations of the data for analysis purposes and to identify errors or shortcomings in the data that can lead to subsequent improvements.
- Substantive research by analysts on an agency's staff will have credibility because of the agency's commitment to openness about the data provided and maintaining independence from political control.
- Substantive research by analysts on an agency's staff can assist in formulating the agency's data program, suggesting changes in priorities, concepts, and needs for new data or discontinuance of outmoded or little-used series.

As with descriptive analyses provided by the agency, substantive analyses should be designed to be relevant to policy by addressing topics of public interest and concern. However, such analyses should not include positions on policy options or be designed to reflect any particular policy agenda. These issues are discussed in Martin (1981), Norwood (1975), and Triplett (1991).

Research on Methodology and Operations

For statistical agencies to be innovative in methods for data collection, analysis, and dissemination, research on methodology and operational procedures must be ongoing. Methodological research may be directed toward improving survey design, measuring error and, when possible, reducing it from such sources as nonresponse and reporting errors, reducing the time and effort asked of respondents, evaluating the best mix of interview modes (e.g., mail, telephone, personal interview) to cope with increasing nonresponse rates due to such phenomena as cell-phone-only households, developing new and improved summary measures and estimation techniques, and developing innovative statistical methods for confidentiality protection. Research on operational procedures may be directed toward facilitating data collection in the field, improving the efficiency and reproducibility of data capture and processing, and enhancing the usability of Internet-based data dissemination systems.

Much of current practice in statistical agencies was developed through research they conducted or obtained from other agencies. Federal statistical agencies, frequently in partnership with academic researchers, pioneered the applications of statistical probability sampling, the national economic accounts, input-output models, and other analytic methods. The U.S. Census Bureau pioneered the use of computers for processing the census, and research on data collection, processing, and dissemination operations continues to lead to creative uses of automated procedures and equipment in these areas. Several federal statistical agencies sponsor research using academic principles of cognitive psychology to improve the design of questionnaires, the clarity of data presentation, and the ease of use of electronic data collection and dissemination tools such as the Internet. The history of the statistical agencies has shown repeatedly that methodological and operations research can lead to large productivity gains in statistical activities at relatively low cost.

An effective statistical agency actively partners with the academic community for methodological research. It also seeks out academic and industry expertise for improving data collection, processing, and dissemination operations. For example, a statistical agency can learn techniques and best practices for improving software development processes from computer scientists (see National Research Council, 2003e, 2004d).

Research on Policy Uses

Much more needs to be known about how statistics are actually used in the policy-making process, both inside and outside the government. Research about how the information produced by a statistical agency is used in practice should contribute to future improvements in design, concepts, and format of data products. For example, public-use files of statistical microdata were developed in response to the growing analytic needs of government and academic researchers.

Gaining an understanding of the variety of uses and users of an agency's data is only a first step. More in-depth research on the policy uses of an agency's information might, for example, explore the use of data in microsimulation or other economic models, or go further to examine how the information from such models and other sources is used in decision making (see National Research Council, 1991a, 1991b, 1997a, 2000b, 2001b, 2003a).

Practice 9: Professional Advancement of Staff

An effective federal statistical agency has personnel policies that encourage the development and retention of a strong professional staff who are committed to the highest standards of quality work. There are several key elements of such a policy:

- The required levels of technical and professional qualifications for positions in the agency are identified, and the agency adheres to these requirements in recruitment and professional development of staff. Position requirements take account of the different kinds of technical and other skills, such as supervisory skills, that are necessary for an agency to have a full range of qualified staff, including not only statisticians, but also experts in relevant subject-matter areas, data collection, processing, and dissemination processes, and management of complex, technical operations.

• Continuing technical education and training, appropriate to the needs of their positions, is provided to staff through in-house training programs and opportunities for external education and training.

• Position responsibilities are structured to ensure that staff have the opportunity to participate, in ways appropriate to their experience and expertise, in research and development activities to improve data quality and cost-effectiveness of agency operations.

• Professional activities, such as publishing in refereed journals and presentations at conferences, are encouraged and recognized, including presentations of technical work in progress with appropriate disclaimers. Participation in relevant statistical and other scientific associations is encouraged to promote interactions with researchers and methodologists in other organizations. Such participation is also a mechanism for openness about the data provided.

• Interaction with other professionals is increased through technical advisory committees, supervision of contract research and research consultants, fellowship programs of visiting researchers, exchange of staff with relevant statistical, policy, or research organizations, and opportunities for new assignments within the agency.

• Accomplishment is rewarded by appropriate recognition and by affording opportunity for further professional development. The prestige and credibility of a statistical agency is enhanced by the professional visibility of its staff, which may include establishing high-level nonmanagement positions for highly qualified technical experts.

An effective statistical agency considers carefully the costs and benefits—monetary and nonmonetary—of using contractor organizations, not only for data collection as most agencies do, but also to supplement in-house staff in other areas.[8] Outsourcing can have benefits, such as: providing experts in areas in which the agency is unlikely to be able to attract highly qualified in-house staff (e.g., some information technology functions), enabling an agency to handle an increase in its workload that is expected to be temporary or that requires specialized skills, and allowing an agency to learn from best industry practices. However, outsourcing can also have costs, including that agency staff become primarily contract managers

[8]Only the Bureau of Labor Statistics and the Census Bureau maintain their own interviewing staff.

and less qualified as technical experts and leaders in their fields. An effective statistical agency maintains and develops a sufficiently large number of in-house staff, including mathematical statisticians, who are qualified to analyze the agency's data and to plan, design, carry out, and evaluate its core operations so that the agency maintains the integrity of its data and its credibility in planning and fulfilling its mission. Statistical agencies should also maintain and develop staff with the expertise necessary for effective management of contractor resources.

An effective statistical agency has policies and practices to instill the highest possible commitment to professional ethics among its staff, as well as procedures for monitoring contractor compliance with ethical standards. When an agency comes under pressure to act against its principles—for example, if it is asked to disclose confidential information for an enforcement purpose or to support an inaccurate interpretation of its data—it must be able to rely on its staff to resist such actions as contrary to the ethical principles of their profession. An effective agency refers its staff to such statements of professional practice as the guidelines published by the American Statistical Association (1999) and the International Statistical Institute (1985), as well as to the agency's own statements about protection of confidentiality, respect for privacy, standards for data quality, and similar matters. It endeavors in other ways to ensure that its staff are fully cognizant of the ethics that must guide their actions in order for the agency to maintain its credibility as a source of objective, reliable information for use by all.

Practice 10: A Strong Internal and External Evaluation Program

Statistical agencies that fully follow such practices as continual development of more useful data, openness about sources and limitations of the data provided, wide dissemination of data, commitment to quality and professional standards of practice, and an active research program will likely be in a good position to make continuous assessments of and improvements in the relevance and quality of their data collection systems. Yet even the best functioning agencies will benefit from an explicit program of internal and independent external evaluations, which frequently offer fresh perspectives. Such evaluations need to address not only specific agency programs, but also the agency's portfolio of programs considered as a whole.

Evaluating Quality

Evaluation of data quality for a continuing survey or other kind of data collection program begins with regular monitoring of quality indicators. For surveys, such monitoring includes unit and item response rates, population coverage rates, and information on sampling error, such as coefficients of variation. (The American Community Survey provides these indicators on its web page, http://www.census.gov/acs/www/ [December 2008].) In addition, in-depth assessment of quality on a wide range of dimensions—including sampling and nonsampling errors across time and among population groups and geographic areas—needs to be undertaken on a periodic basis (National Research Council, 2007b).

Research on methods to improve data quality may cover such areas as alternative methods for imputing values for missing data and alternative question designs, using cognitive methods, to reduce respondent reporting errors. Methods for such research may include the use of "methods panels" (small samples of respondents with whom experiments are conducted by using alternative procedures and questionnaires), matching with administrative records, simulations of sensitivity to alternative procedures, and the like. The goal of the research is the development of feasible, cost-effective improved procedures for implementation.

In ongoing programs for which it is disruptive to implement improvements on a continuing basis, a common practice is to undertake major research and development activities at intervals of, say, 5 or 10 years or longer. Agencies should ensure, however, that the intervals between major research and development activities do not become so long that data collection programs deteriorate in quality, relevance, and efficiency over time.

Regular, well-designed program evaluations, with adequate budget support, are key to ensuring that data collection programs do not deteriorate. Having a set schedule for research and development efforts will enable data collection managers to ensure that the quality and usefulness of their data are maintained and help prevent the locking into place of increasingly less optimal procedures over time.

Evaluating Relevance

In addition to quality, it is important to assess the relevance of an agency's data collection programs. The question in this instance is whether the agency is "doing the right thing" in contrast to whether the agency is "doing things right." Relevance should be assessed not only for particular programs or closely related sets of programs, but also for an agency's complete portfolio to assist it in making the best choices among program priorities given the available resources.

Keeping in close touch with stakeholders and important user constituencies—through such means as regular meetings, workshops, conferences, and other activities—is important to ensuring relevance. Customer surveys can be helpful on some aspects of relevance, although they typically provide only gross indicators of customer satisfaction, usually with regard to timeliness and ease of use of data products. As discussed in the next section, including other federal statistical colleagues, both as users and as collaborators, in this communication can also be valuable.

Statistical agencies commonly find that it is difficult to discontinue or scale back a particular data series, even when it has largely outlived its usefulness relative to other series, because of objections by users who have become accustomed to it. In the face of limited resources, however, discontinuing a series is preferable to across-the-board cuts in all programs that reduce the accuracy and usefulness of the more relevant and less relevant data series alike. Regular internal and external reviews can help an agency not only reassess its priorities, but also develop the justification and support for changes to its portfolio.

Types of Reviews

Regular program reviews should include a mixture of internal and external evaluation. Agency staff should set goals and timetables for internal evaluations, which should involve staff who do not regularly work on the program under review. Independent external evaluations should also be conducted on a regular basis, the frequency of which should depend on the importance of the data and on how quickly changes in such factors as respondent behavior and data collection technology may adversely affect a program. In a world in which people and organizations appear increasingly less willing to respond to surveys, it becomes urgent to continually monitor response and have more frequent evaluations than in a more stable environment. In addition to program evaluations, agencies should seek outside reviews to examine priorities and quality practices across the entire agency.

External reviews can take many forms. They may include recommendations from advisory committees that meet at regular intervals (typically every 6 months). However, advisory committees should never be the sole source of outside review because the members of such committees rarely have the opportunity to become deeply familiar with agency programs. Administrations often develop evaluation mechanisms (see Appendix B) that may be helpful to an agency. External reviews can also take the form of a "visiting committee" using the National Science Foundation model or academic models (see, e.g., http://www.nsf.gov/od/oia/activities/cov/covs.jsp [December 2008]); or a special committee established by a relevant professional association (see, e.g., American Statistical Association, 1984); or a study by a panel of experts (see, e.g., National Research Council, 1985a, 1985b, 1986, 1993a, 1997b, 2000b, 2000c, 2003c, 2004c, 2004d, 2008c, 2008d).

Practice 11: Coordination and Cooperation with Other Statistical Agencies

The U.S. federal statistical system consists of many agencies in different departments, each with its own mission. Nonetheless, statistical agencies do not and should not conduct their activities in isolation. An effective statistical agency actively explores ways to work with other agencies to meet current information needs, for example, by seeking ways to integrate the designs of existing data systems to provide new or more useful data than a single system can provide. An effective agency is also alert for occasions when it can provide technical assistance to other agencies—including not only other statistical agencies, but also program agencies in its department—as well as occasions when it can receive such assistance in turn. Efforts to standardize concepts and definitions, such as those for industries, occupations, and race and ethnicity, further contribute to effective coordination of statistical agency endeavors, as does the development of broad macro models, such as the system of national accounts (see, e.g., National Research Council, 2004a, 2004b; also see Appendix B). Initiatives for sharing data among statistical agencies (including individual data and address lists when permitted by law and when sharing does not violate confidentiality promises) can be helpful for such purposes as achieving greater efficiency in drawing samples, evaluating completeness of population coverage, and reducing duplication among statistical programs, as well as reducing respondent burden.

The responsibility for coordinating statistical work in the federal government is specifically assigned to the Office of Information and Regulatory Affairs (OIRA) in OMB by the Paperwork Reduction Act (previously, by the Federal Reports Act and the Budget and Accounting Procedures Act—

see Appendix B). The Statistical and Science Policy Office in OIRA, often working with the assistance of interagency committees, reviews concepts of interest to more than one agency; issues standard classification systems (of industries, metropolitan areas, etc.) and oversees their periodic revision; consults with other parts of OMB on statistical budgets; and, by reviewing statistical information collections as well as the statistical programs of the government as a whole, identifies gaps in statistical data, programs that may be duplicative, and areas in which interagency cooperation might lead to greater efficiency and added utility of data. The Statistical and Science Policy Office also is responsible for coordinating U.S. participation in international statistical activities.[9]

The Statistical and Science Policy Office encourages the use of administrative data for statistical purposes, when feasible, and works to establish common goals and norms on major statistical issues, such as confidentiality. It sponsors and heads the interagency Federal Committee on Statistical Methodology (FCSM), which issues guidelines and recommendations on statistical issues common to a number of agencies (see Federal Committee on Statistical Methodology, 1978a-2005; for the papers from the FCSM 2007 research conference, see http://www.fcsm.gov/events/papers2007 [December 2008]). It encourages the Committee on National Statistics at the National Academies to serve as an independent adviser and reviewer of federal statistical activities. The 1995 reauthorization of the Paperwork Reduction Act created a statutory basis for the Interagency Council on Statistical Policy (ICSP), formalizing an arrangement whereby statistical agency heads participated with OMB in activities to coordinate federal statistical programs (see Appendixes A and B).

There are many forms of interagency cooperation and coordination. Some efforts are multilateral, some bilateral. Many result from common interests in specific subject areas, such as economic statistics, statistics on people with disabilities, or statistics on children or the elderly. U.S. Office of Management and Budget (2008c:Ch. 3) describes several interagency collaborative efforts, such as joint support for research that fosters new and innovative approaches to surveys, expansion and improvement of the coverage and features of *FedStats*, which provides access to statistics from

---

[9]The Statistical and Science Policy Office was renamed from the Statistical Policy Office to reflect added responsibilities with respect to the 2001 Information Quality Act standards and guidelines, OMB's guidance on peer review planning and implementation, and evaluations of science underlying proposed regulatory actions.

more than 100 government agencies at http://www.fedstats.gov [December 2008], and implementation of comparable measures of disability on major household surveys.

A common type of bilateral arrangement is the agreement of a program agency to provide administrative data to a statistical agency to be used as a sampling frame, a source of classification information, or a summary compilation to check (and possibly revise) preliminary sample results. The Bureau of Labor Statistics, for example, benchmarks its monthly establishment employment reports to data supplied by state employment security agencies. Such practices improve statistical estimates, reduce costs, and eliminate duplicate requests for information from the same respondents. In other cases, federal statistical agencies engage in cooperative data collection with state counterparts to let one collection system satisfy the needs of both. A number of such joint systems have been developed, notably by the Bureau of Labor Statistics, the National Agricultural Statistics Service, the National Center for Education Statistics, and the National Center for Health Statistics.

Another example of a joint arrangement is the case in which one statistical agency contracts with another to conduct a survey, compile special tabulations, or develop models. Such arrangements make use of the special skills of the supplying agency and facilitate use of common concepts and methods. The Census Bureau conducts many surveys for other agencies, both the National Center for Health Statistics and the National Agricultural Statistics Service receive funding from other agencies in their departments to support their survey work, and the Division of Science Resources Statistics receives funding from agencies in other departments to support several of its surveys (see U.S. Office of Management and Budget, 2008c:Table 2).

The major federal statistics agencies are also concerned with international comparability of statistics. Under the leadership of OMB's Statistical and Science Policy Office, they contribute to the deliberations of the United Nations Statistical Commission, the Organisation for Economic Co-operation and Development, and other international organizations, participate in the development of international standard classifications and systems, and support educational activities that promote improved statistics in developing countries. Statistical agencies also learn from and contribute to the work of established statistical agencies in other countries in such areas as survey methodology, record linkage, confidentiality protection techniques, and data quality standards. Several statistical agencies run educational programs for government statisticians in developing countries.

Some statistical agencies have long-term cooperative relationships with international groups, for example, the Bureau of Labor Statistics with the International Labor Organization, the National Agricultural Statistics Service with the Food and Agriculture Organization, the National Center for Education Statistics with the International Indicators of Education Systems project of the Organisation for Economic Co-operation and Development, and the National Center for Health Statistics with the World Health Organization.

To be of most value, the efforts of statistical agencies to cooperate as partners with one another should involve the full range of their activities, including definitions, concepts, measurement methods, analytical tools, dissemination modes, and disclosure limitation techniques. Such efforts should also extend to policies and professional practices, so that agencies can respond effectively and with a coordinated voice to such government-wide initiatives as data quality guidelines, privacy impact assessments, performance rating criteria, institutional review board requirements, and others.

Finally, coordination efforts should encompass the development of data, especially for emerging policy issues (National Research Council, 1999a). In some cases, it may be not only more efficient, but also productive of needed new data for agencies to fully integrate the designs of existing data systems, such as when one survey provides the sampling frame for a related survey. In other instances, cooperative efforts may identify ways for agencies to improve their individual data systems so that they are more useful for a wide range of purposes.

Two of the more effective continuing cooperative efforts in this regard have been the Federal Interagency Forum on Aging-Related Statistics and the Federal Interagency Forum on Child and Family Statistics. The former was established in the mid-1980s by the National Institute on Aging, in cooperation with the National Center for Health Statistics and the Census Bureau. The forum's goals include coordinating the development and use of statistical data bases among federal agencies, identifying information gaps and data inconsistencies, and encouraging cross-national research and data collection for the aging population. The forum was reorganized in 1998 to include six new member agencies and has grown over the years to include 15 agencies. The forum develops a periodic indicators chart book, which was first published in 2000 and was most recently issued in 2008 (Federal Interagency Forum on Aging-Related Statistics, 2008).

The Federal Interagency Forum on Child and Family Statistics was formalized in a 1994 executive order to foster coordination and collaboration in the collection and reporting of federal data on children and families. Its membership currently includes 22 statistical and program agencies. The forum's reports (e.g., Federal Interagency Forum on Child and Family Statistics, 2007, 2008) describe the condition of America's children, including changing population and family characteristics, the environment in which children are living, and indicators of well-being in the areas of economic security, health, behavior, social environment, and education.

No single agency, whether a statistical or program agency, could have produced the forum reports alone. Working together in this way, federal statistical agencies contribute to presenting data in a form that is more relevant to policy concerns and to a stronger statistical system overall.

References

American Statistical Association (1984). Report of the ASA technical panel on the census undercount. *American Statistician* 38(4):252-256.

_______ (1999). *Ethical Guidelines for Statistical Practice.* Alexandria, VA. Available: www.amstat.org/profession/index.cfm?fuseaction=main.

Anderson, Margo J. (1988). *The American Census: A Social History.* New Haven, CT: Yale University Press.

Brackstone, Gordon (1999). Managing data quality in a statistical agency. *Survey Methodology* 25(2):139-149.

Chakrabarty, Rameswar, assisted by Georgina Torres (1996). *The American Housing Survey—A Quality Profile.* Current Housing Reports, H121/95-1. Office of Policy Development and Research, U.S. Department of Housing and Urban Development and U.S. Census Bureau. Washington, DC: U.S. Department of Commerce.

de Leeuw, Edith, and Wim de Heer (2002). Trends in household survey nonresponse: A longitudinal and international comparison. Pp. 41-54 in *Survey Nonresponse,* Robert M. Groves, Donald A. Dillman, John L. Eltinge, and Roderick J.A. Little, eds. New York: John Wiley & Sons.

Doyle, Pat, Julia I. Lane, Jules J.M. Theeuwes, and Laura V. Zayatz, eds. (2001). *Confidentiality, Disclosure, and Data Access: Theory and Practical Applications for Statistical Agencies.* Amsterdam: Elsevier North-Holland.

Duncan, Joseph W., and William C. Shelton (1978). *Revolution in United States Government Statistics, 1926-1976.* Office of Federal Statistical Policy and Standards. Washington, DC: U.S. Department of Commerce.

Energy Information Administration (1996). *Residential Energy Consumption Survey Quality Profile.* Prepared by Thomas B. Jabine. Washington, DC: U.S. Department of Energy.

Federal Committee on Statistical Methodology (1978a). *Report on Statistics for Allocation of Funds.* Statistical Policy Working Paper 1 (NTIS PB86-211521/AS). Washington, DC: U.S. Department of Commerce.

_______ (1978b). *Report on Statistical Disclosure and Disclosure-Avoidance Techniques.* Statistical Policy Working Paper 2 (NTIS PB86-211539/AS). Washington, DC: U.S. Department of Commerce.

_______ (1978c). *An Error Profile: Employment as Measured by the Current Population Survey.* Statistical Policy Working Paper 3 (NTIS PB86-214269/AS). Washington, DC: U.S. Department of Commerce.

_______ (1978d). *Glossary of Nonsampling Error Terms: An Illustration of a Semantic Problem in Statistics.* Statistical Policy Working Paper 4 (NTIS PB86-211547/AS). Washington, DC: U.S. Department of Commerce.

_______ (1980a). *Report on Exact and Statistical Matching Techniques.* Statistical Policy Working Paper 5 (NTIS PB86-215829/AS). Washington, DC: U.S. Department of Commerce.

_______ (1980b). *Report on Statistical Uses of Administrative Records.* Statistical Policy Working Paper 6 (NTIS PB86-214285/AS). Washington, DC: U.S. Department of Commerce.

_______ (1982a). *An Interagency Review of Time-Series Revision Policies.* Statistical Policy Working Paper 7 (NTIS PB86-232451/AS). Washington, DC: U.S. Office of Management and Budget.

_______ (1982b). *Statistical Interagency Agreements.* Statistical Policy Working Paper 8 (NTIS PB86-230570/AS). Washington, DC: U.S. Office of Management and Budget.

_______ (1983a). *Contracting for Surveys.* Statistical Policy Working Paper 9 (NTIS PB83-233148). Washington, DC: U.S. Office of Management and Budget.

_______ (1983b). *Approaches to Developing Questionnaires.* Statistical Policy Working Paper 10 (NTIS PB84-105055). Washington, DC: U.S. Office of Management and Budget.

_______ (1984a). *A Review of Industry Coding Systems.* Statistical Policy Working Paper 11 (NTIS PB84-135276). Washington, DC: U.S. Office of Management and Budget.

_______ (1984b). *The Role of Telephone Data Collection in Federal Statistics.* Statistical Policy Working Paper 12 (NTIS PB85-105971). Washington, DC: U.S. Office of Management and Budget.

_______ (1986). *Federal Longitudinal Surveys.* Statistical Policy Working Paper 13 (NTIS PB86-139730). Washington, DC: U.S. Office of Management and Budget.

_______ (1987). *Workshop on Statistical Uses of Microcomputer in Federal Agencies.* Statistical Policy Working Paper 14 (NTIS PB87-166393). Washington, DC: U.S. Office of Management and Budget.

_______ (1988). *Quality in Establishment Surveys.* Statistical Policy Working Paper 15 (NTIS PB88-232921). Washington, DC: U.S. Office of Management and Budget.

_______ (1990a). *A Comparative Study of Reporting Units in Selected Employer Data Systems.* Statistical Policy Working Paper 16 (NTIS PB90-205238). Washington, DC: U.S. Office of Management and Budget.

_______ (1990b). *Survey Coverage.* Statistical Policy Working Paper 17 (NTIS PB90-205246). Washington, DC: U.S. Office of Management and Budget.

_______ (1990c). *Data Editing in Federal Statistical Agencies.* Statistical Policy Working Paper 18 (NTIS PB90-205253). Washington, DC: U.S. Office of Management and Budget.

_______ (1990d). *Computer Assisted Survey Information.* Statistical Policy Working Paper 19 (NTIS PB90-205261). Washington, DC: U.S. Office of Management and Budget.

_______ (1991). *Seminar on Quality of Federal Data.* Statistical Policy Working Paper 20 (NTIS PB91-142414). Washington, DC: U.S. Office of Management and Budget.

_______ (1993). *Indirect Estimators in Federal Programs.* Statistical Policy Working Paper 21 (NTIS PB93-209294). Washington, DC: U.S. Office of Management and Budget.

_______ (1994). *Report on Statistical Disclosure Limitation Methodology.* Statistical Policy Working Paper 22 (NTIS PB94-165305). Washington, DC: U.S. Office of Management and Budget. [revised in 2005]

_______ (1995a). *Seminar on New Directions in Statistical Methodology.* Statistical Policy Working Paper 23 (NTIS PB95-182978). Washington, DC: U.S. Office of Management and Budget.

_______ (1995b). *Electronic Dissemination of Statistical Data.* Statistical Policy Working Paper 24 (NTIS PB96-121629). Washington, DC: U.S. Office of Management and Budget.

_______ (1996). *Data Editing Workshop and Exposition.* Statistical Policy Working Paper 25 (NTIS PB97-104624). Washington, DC: U.S. Office of Management and Budget.

_______ (1997). *Seminar on Statistical Methodology in the Public Service.* Statistical Policy Working Paper 26 (NTIS PB97-162580). Washington, DC: U.S. Office of Management and Budget.

_______ (1998). *Training for the Future: Addressing Tomorrow's Survey Tasks.* Statistical Policy Working Paper 27 (NTIS PB99-102576). Washington, DC: U.S. Office of Management and Budget.

_______ (1999a). *Seminar on Interagency Coordination and Cooperation.* Statistical Policy Working Paper 28 (NTIS PB99-132029). Washington, DC: U.S. Office of Management and Budget.

_______ (1999b). *Federal Committee on Statistical Methodology Research Conference (Conference Papers).* Statistical Policy Working Paper 29 (NTIS PB99-166795). Washington, DC: U.S. Office of Management and Budget.

_______ (1999c). *1999 Federal Committee on Statistical Methodology Research Conference: Complete Proceedings.* Statistical Policy Working Paper 30 (NTIS PB2000-105886). Washington, DC: U.S. Office of Management and Budget.

_______ (2001a). *Measuring and Reporting Sources of Error in Surveys.* Statistical Policy Working Paper 31 (NTIS PB2001-104329). Washington, DC: U.S. Office of Management and Budget.

_______ (2001b). *Seminar on Integrating Federal Statistical Information and Processes.* Statistical Policy Working Paper 32 (NTIS PB2001-104626). Washington, DC: U.S. Office of Management and Budget.

_______ (2001c). *Seminar on the Funding Opportunity in Survey Research.* Statistical Policy Working Paper 33 (NTIS PB2001-108851). Washington, DC: U.S. Office of Management and Budget.

_______ (2001d). *Federal Committee on Statistical Methodology Research Conference.* Statistical Policy Working Paper 34 (NTIS PB2002-100103). Washington, DC: U.S. Office of Management and Budget.

_______ (2003a). *Seminar on the Funding Opportunity in Survey Research.* Statistical Policy Working Paper 36. Washington, DC: U.S. Office of Management and Budget.

_______ (2003b). *Federal Committee on Statistical Methodology Research Conference.* Statistical Policy Working Paper 37. Washington, DC: U.S. Office of Management and Budget.

_______ (2004a). *Seminar on Challenges to the Federal Statistical System in Fostering Access to Statistics.* Statistical Policy Working Paper 35. Washington, DC: U.S. Office of Management and Budget.

_______ (2004b). *Summary Report of the FCSM-GSS Workshop on Web-Based Data Collection.* Statistical Policy Working Paper 38. Washington, DC: U.S. Office of Management and Budget.

_______ (2005). *Federal Committee on Statistical Methodology Research Conference, Conference Papers.* Washington, DC: U.S. Office of Management and Budget.

Federal Interagency Forum on Aging-Related Statistics (2008). *Older Americans 2008: Key Indicators of Well-Being.* Available: www.agingstats.gov.

Federal Interagency Forum on Child and Family Statistics (2007). *America's Children: Key National Indicators of Well-Being, 2007.* Washington, DC: U.S. Government Printing Office. Available: www.childstats.gov.

_______ (2008). *America's Children in Brief: Key National Indicators of Well-Being, 2008.* Washington, DC: U.S. Government Printing Office. Available: www.childstats.gov.

Gates, Gerald W. (2000). Confidentiality. Pp. 80-83 in Margo J. Anderson, ed. *Encyclopedia of the U.S. Census.* Washington, DC: CQ Press.

Hillygus, D. Sunshine, Norman H. Nie, Kenneth Prewitt, and Heili Pals (2006). *The Hard Count: The Political and Social Challenges of Census Mobilization.* New York: Russell Sage Foundation.

International Statistical Institute (1985). *Declaration on Professional Ethics.* Voorburg, The Netherlands. Available: www.isi.cbs.nl/ethics.htm.

Kalton, Graham, Marianne Winglee, Sheila Krawchuk, and Daniel Levine (2000). *Quality Profile for SASS Rounds 1-3: 1987-1995, Aspects of the Quality of Data in the Schools and Staffing Surveys (SASS).* National Center for Education Statistics. Washington, DC: U.S. Department of Education.

Martin, Margaret E. (1981). Statistical practice in bureaucracies. *Journal of the American Statistical Association* 76(373):1-8.

National Research Council (1976). *Setting Statistical Priorities.* Panel on Methodology for Statistical Priorities. Committee on National Statistics. Washington, DC: National Academy Press.

_______ (1984). *Cognitive Aspects of Survey Methodology: Building a Bridge Between Disciplines.* Report of the Advanced Research Seminar on Cognitive Aspects of Survey Methodology, Thomas B. Jabine, Miron L. Straf, Judith M. Tanur, and Roger Tourangeau, eds. Committee on National Statistics. Washington, DC: National Academy Press.

_______ (1985a). *Immigration Statistics: A Story of Neglect.* Panel on Immigration Statistics, Daniel B. Levine, Kenneth Hill, and Robert Warren, eds. Committee on National Statistics. Washington, DC: National Academy Press.

_______ (1985b). *Natural Gas Data Needs in a Changing Regulatory Environment.* Panel on Natural Gas Statistics. Committee on National Statistics. Washington, DC: National Academy Press.

_______ (1985c). *Sharing Research Data.* Stephen E. Fienberg, Margaret E. Martin, and Miron L. Straf, eds. Committee on National Statistics. Washington, DC: National Academy Press.

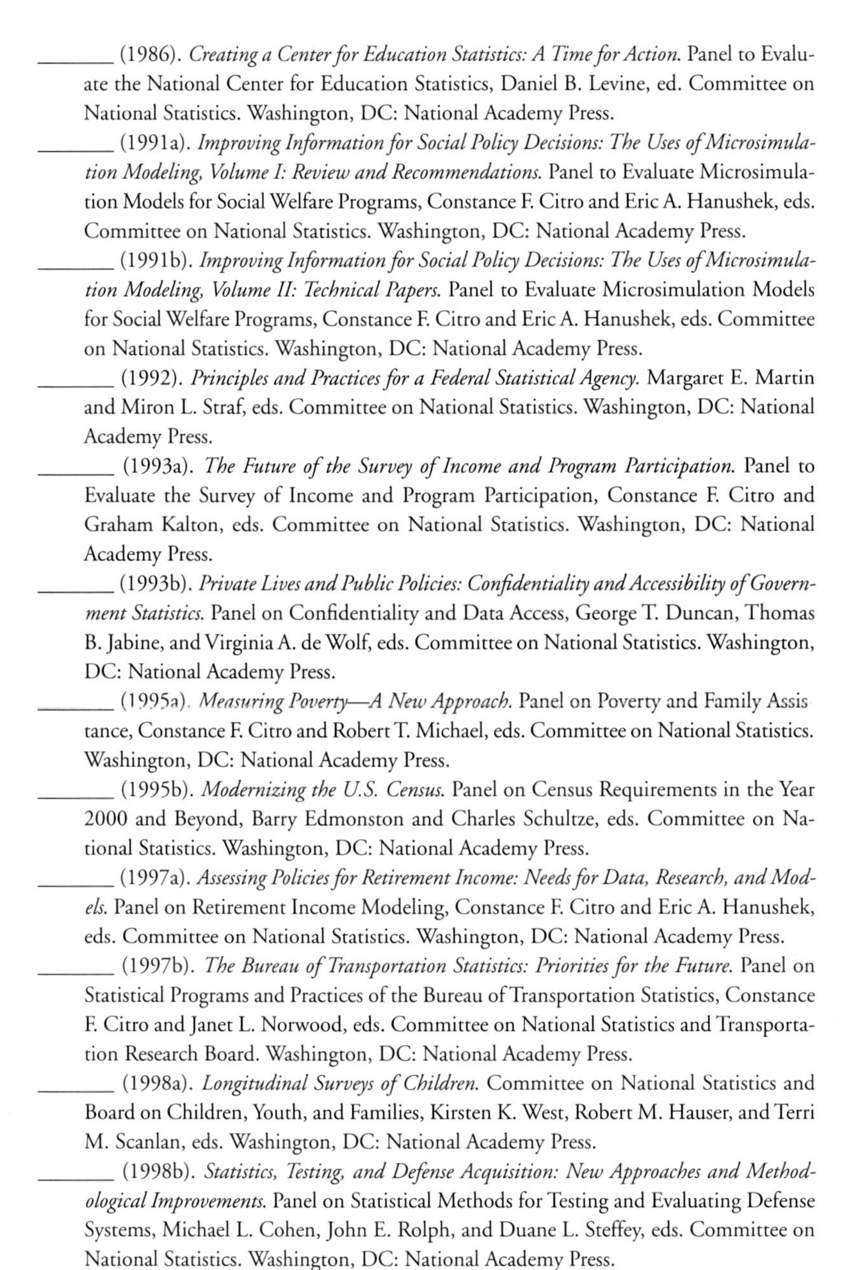

_______ (1986). *Creating a Center for Education Statistics: A Time for Action.* Panel to Evaluate the National Center for Education Statistics, Daniel B. Levine, ed. Committee on National Statistics. Washington, DC: National Academy Press.

_______ (1991a). *Improving Information for Social Policy Decisions: The Uses of Microsimulation Modeling, Volume I: Review and Recommendations.* Panel to Evaluate Microsimulation Models for Social Welfare Programs, Constance F. Citro and Eric A. Hanushek, eds. Committee on National Statistics. Washington, DC: National Academy Press.

_______ (1991b). *Improving Information for Social Policy Decisions: The Uses of Microsimulation Modeling, Volume II: Technical Papers.* Panel to Evaluate Microsimulation Models for Social Welfare Programs, Constance F. Citro and Eric A. Hanushek, eds. Committee on National Statistics. Washington, DC: National Academy Press.

_______ (1992). *Principles and Practices for a Federal Statistical Agency.* Margaret E. Martin and Miron L. Straf, eds. Committee on National Statistics. Washington, DC: National Academy Press.

_______ (1993a). *The Future of the Survey of Income and Program Participation.* Panel to Evaluate the Survey of Income and Program Participation, Constance F. Citro and Graham Kalton, eds. Committee on National Statistics. Washington, DC: National Academy Press.

_______ (1993b). *Private Lives and Public Policies: Confidentiality and Accessibility of Government Statistics.* Panel on Confidentiality and Data Access, George T. Duncan, Thomas B. Jabine, and Virginia A. de Wolf, eds. Committee on National Statistics. Washington, DC: National Academy Press.

_______ (1995a). *Measuring Poverty—A New Approach.* Panel on Poverty and Family Assistance, Constance F. Citro and Robert T. Michael, eds. Committee on National Statistics. Washington, DC: National Academy Press.

_______ (1995b). *Modernizing the U.S. Census.* Panel on Census Requirements in the Year 2000 and Beyond, Barry Edmonston and Charles Schultze, eds. Committee on National Statistics. Washington, DC: National Academy Press.

_______ (1997a). *Assessing Policies for Retirement Income: Needs for Data, Research, and Models.* Panel on Retirement Income Modeling, Constance F. Citro and Eric A. Hanushek, eds. Committee on National Statistics. Washington, DC: National Academy Press.

_______ (1997b). *The Bureau of Transportation Statistics: Priorities for the Future.* Panel on Statistical Programs and Practices of the Bureau of Transportation Statistics, Constance F. Citro and Janet L. Norwood, eds. Committee on National Statistics and Transportation Research Board. Washington, DC: National Academy Press.

_______ (1998a). *Longitudinal Surveys of Children.* Committee on National Statistics and Board on Children, Youth, and Families, Kirsten K. West, Robert M. Hauser, and Terri M. Scanlan, eds. Washington, DC: National Academy Press.

_______ (1998b). *Statistics, Testing, and Defense Acquisition: New Approaches and Methodological Improvements.* Panel on Statistical Methods for Testing and Evaluating Defense Systems, Michael L. Cohen, John E. Rolph, and Duane L. Steffey, eds. Committee on National Statistics. Washington, DC: National Academy Press.

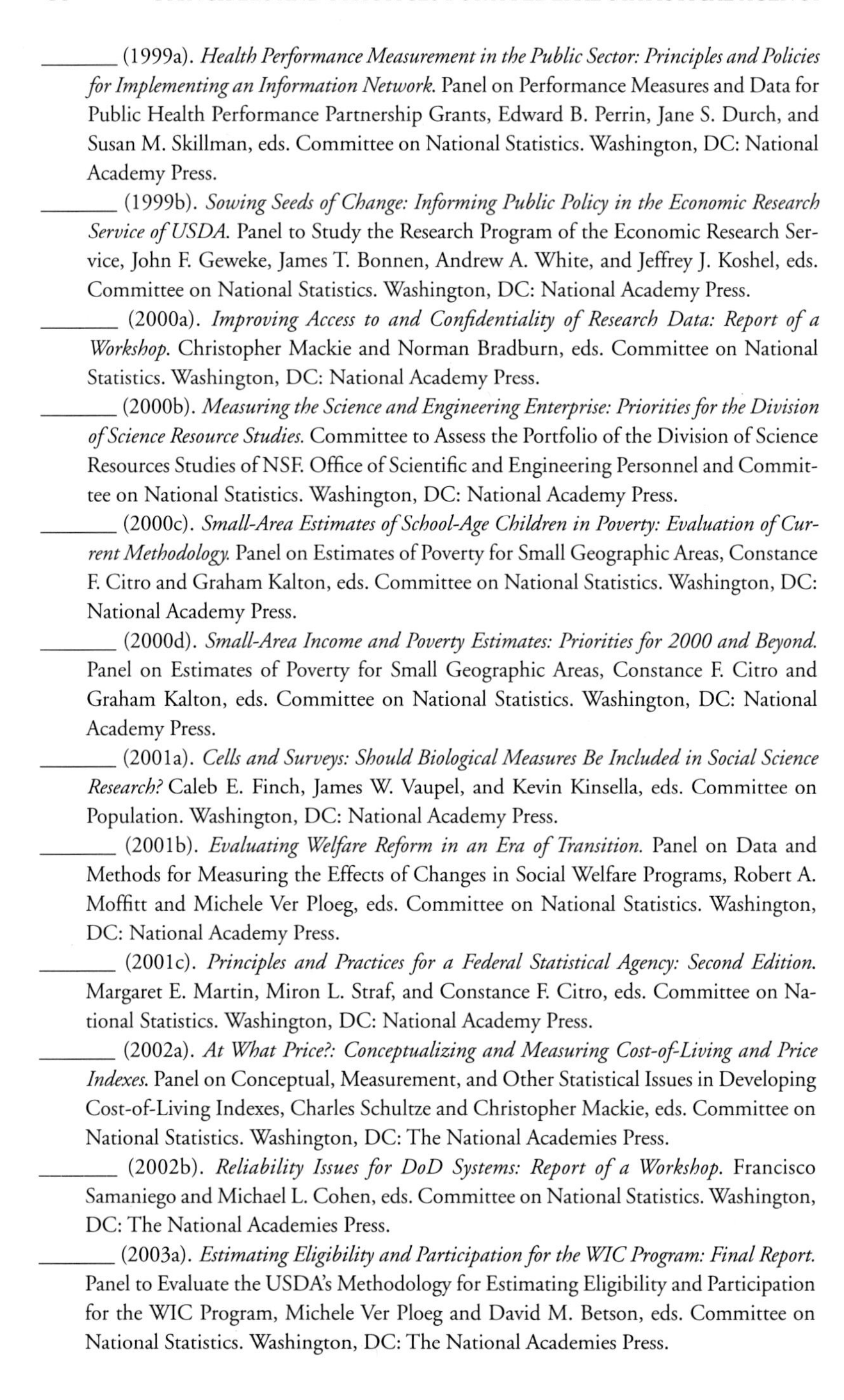

_______ (1999a). *Health Performance Measurement in the Public Sector: Principles and Policies for Implementing an Information Network.* Panel on Performance Measures and Data for Public Health Performance Partnership Grants, Edward B. Perrin, Jane S. Durch, and Susan M. Skillman, eds. Committee on National Statistics. Washington, DC: National Academy Press.

_______ (1999b). *Sowing Seeds of Change: Informing Public Policy in the Economic Research Service of USDA.* Panel to Study the Research Program of the Economic Research Service, John F. Geweke, James T. Bonnen, Andrew A. White, and Jeffrey J. Koshel, eds. Committee on National Statistics. Washington, DC: National Academy Press.

_______ (2000a). *Improving Access to and Confidentiality of Research Data: Report of a Workshop.* Christopher Mackie and Norman Bradburn, eds. Committee on National Statistics. Washington, DC: National Academy Press.

_______ (2000b). *Measuring the Science and Engineering Enterprise: Priorities for the Division of Science Resource Studies.* Committee to Assess the Portfolio of the Division of Science Resources Studies of NSF. Office of Scientific and Engineering Personnel and Committee on National Statistics. Washington, DC: National Academy Press.

_______ (2000c). *Small-Area Estimates of School-Age Children in Poverty: Evaluation of Current Methodology.* Panel on Estimates of Poverty for Small Geographic Areas, Constance F. Citro and Graham Kalton, eds. Committee on National Statistics. Washington, DC: National Academy Press.

_______ (2000d). *Small-Area Income and Poverty Estimates: Priorities for 2000 and Beyond.* Panel on Estimates of Poverty for Small Geographic Areas, Constance F. Citro and Graham Kalton, eds. Committee on National Statistics. Washington, DC: National Academy Press.

_______ (2001a). *Cells and Surveys: Should Biological Measures Be Included in Social Science Research?* Caleb E. Finch, James W. Vaupel, and Kevin Kinsella, eds. Committee on Population. Washington, DC: National Academy Press.

_______ (2001b). *Evaluating Welfare Reform in an Era of Transition.* Panel on Data and Methods for Measuring the Effects of Changes in Social Welfare Programs, Robert A. Moffitt and Michele Ver Ploeg, eds. Committee on National Statistics. Washington, DC: National Academy Press.

_______ (2001c). *Principles and Practices for a Federal Statistical Agency: Second Edition.* Margaret E. Martin, Miron L. Straf, and Constance F. Citro, eds. Committee on National Statistics. Washington, DC: National Academy Press.

_______ (2002a). *At What Price?: Conceptualizing and Measuring Cost-of-Living and Price Indexes.* Panel on Conceptual, Measurement, and Other Statistical Issues in Developing Cost-of-Living Indexes, Charles Schultze and Christopher Mackie, eds. Committee on National Statistics. Washington, DC: The National Academies Press.

_______ (2002b). *Reliability Issues for DoD Systems: Report of a Workshop.* Francisco Samaniego and Michael L. Cohen, eds. Committee on National Statistics. Washington, DC: The National Academies Press.

_______ (2003a). *Estimating Eligibility and Participation for the WIC Program: Final Report.* Panel to Evaluate the USDA's Methodology for Estimating Eligibility and Participation for the WIC Program, Michele Ver Ploeg and David M. Betson, eds. Committee on National Statistics. Washington, DC: The National Academies Press.

_______ (2003b). *Innovations in Software Engineering for Defense Systems.* Oversight Committee for the Workshop on Statistical Methods in Software Engineering for Defense Systems, Siddhartha R. Dalal, Jesse H. Poore, and Michael L. Cohen, eds. Committee on National Statistics and Division on Engineering and Physical Sciences. Washington, DC: The National Academies Press.

_______ (2003c). *Measuring Personal Travel and Goods Movement: A Review of the Bureau of Transportation Statistics' Surveys—Special Report 277.* Committee to Review the Bureau of Transportation Statistics' Survey Programs. Committee on National Statistics and the Transportation Research Board. Washington, DC: The National Academies Press.

_______ (2003d). *Protecting Participants and Facilitating Social and Behavioral Sciences Research.* Panel on Institutional Review Boards, Surveys, and Social Science Research, Constance F. Citro, Daniel R. Ilgen, and Cora B. Marrett, eds. Committee on National Statistics and Board on Behavioral, Cognitive, and Sensory Sciences. Washington, DC: The National Academies Press.

_______ (2003e). *Survey Automation: Report and Workshop Proceedings.* Oversight Committee for the Workshop on Survey Automation, Daniel L. Cork, Michael L. Cohen, Robert Groves, and William Kalsbeek, eds. Committee on National Statistics. Washington, DC: The National Academies Press.

_______ (2004a). *Beyond the Market: Designing Nonmarket Accounts for the United States.* Panel to Study the Design of Nonmarket Accounts, Katharine G. Abraham and Christopher Mackie, eds. Committee on National Statistics. Washington, DC: The National Academies Press.

_______ (2004b). *Eliminating Health Disparities: Measurement and Data Needs.* Panel on DHHS Collection of Race and Ethnic Data, Michele Ver Ploeg and Edward Perrin, eds. Committee on National Statistics. Washington, DC: The National Academies Press.

_______ (2004c). *Measuring Research and Development Expenditures in the U.S. Economy.* Panel on Research and Development Statistics at the National Science Foundation, Lawrence D. Brown, Thomas J. Plewes, and Marisa A. Gerstein, eds. Committee on National Statistics. Washington, DC: The National Academies Press.

_______ (2004d). *Reengineering the 2010 Census: Risks and Challenges.* Panel on Research on Future Census Methods, Daniel L. Cork, Michael L. Cohen, and Benjamin F. King, eds. Committee on National Statistics. Washington, DC: The National Academies Press.

_______ (2004e). *The 2000 Census: Counting Under Adversity.* Panel to Review the 2000 Census, Constance F. Citro, Daniel L. Cork, and Janet L. Norwood, eds. Committee on National Statistics. Washington, DC: The National Academies Press.

_______ (2005a). *Beyond the Market: Designing Nonmarket Accounts for the United States.* Panel to Study the Design of Nonmarket Accounts, Katharine G. Abraham and Christopher Mackie, eds. Committee on National Statistics. Washington, DC: The National Academies Press.

_______ (2005b). *Expanding Access to Research Data: Reconciling Risks and Opportunities.* Panel on Data Access for Research Purposes. Committee on National Statistics. Washington, DC: The National Academies Press.

_______ (2005c). *Experimental Poverty Measures: Summary of a Workshop.* John Iceland, rapporteur. Committee on National Statistics. Washington, DC: The National Academies Press.

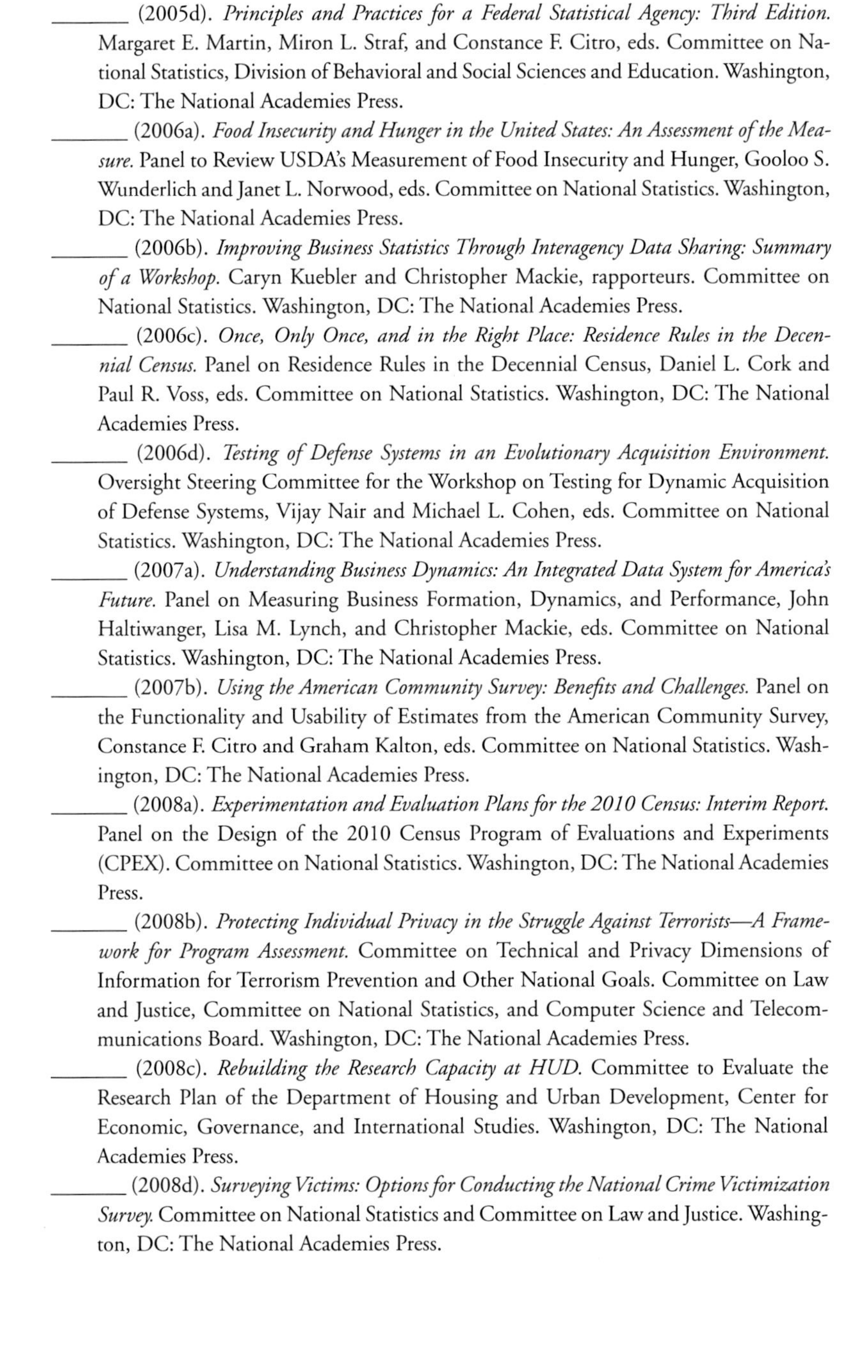

_______ (2005d). *Principles and Practices for a Federal Statistical Agency: Third Edition.* Margaret E. Martin, Miron L. Straf, and Constance F. Citro, eds. Committee on National Statistics, Division of Behavioral and Social Sciences and Education. Washington, DC: The National Academies Press.

_______ (2006a). *Food Insecurity and Hunger in the United States: An Assessment of the Measure.* Panel to Review USDA's Measurement of Food Insecurity and Hunger, Gooloo S. Wunderlich and Janet L. Norwood, eds. Committee on National Statistics. Washington, DC: The National Academies Press.

_______ (2006b). *Improving Business Statistics Through Interagency Data Sharing: Summary of a Workshop.* Caryn Kuebler and Christopher Mackie, rapporteurs. Committee on National Statistics. Washington, DC: The National Academies Press.

_______ (2006c). *Once, Only Once, and in the Right Place: Residence Rules in the Decennial Census.* Panel on Residence Rules in the Decennial Census, Daniel L. Cork and Paul R. Voss, eds. Committee on National Statistics. Washington, DC: The National Academies Press.

_______ (2006d). *Testing of Defense Systems in an Evolutionary Acquisition Environment.* Oversight Steering Committee for the Workshop on Testing for Dynamic Acquisition of Defense Systems, Vijay Nair and Michael L. Cohen, eds. Committee on National Statistics. Washington, DC: The National Academies Press.

_______ (2007a). *Understanding Business Dynamics: An Integrated Data System for America's Future.* Panel on Measuring Business Formation, Dynamics, and Performance, John Haltiwanger, Lisa M. Lynch, and Christopher Mackie, eds. Committee on National Statistics. Washington, DC: The National Academies Press.

_______ (2007b). *Using the American Community Survey: Benefits and Challenges.* Panel on the Functionality and Usability of Estimates from the American Community Survey, Constance F. Citro and Graham Kalton, eds. Committee on National Statistics. Washington, DC: The National Academies Press.

_______ (2008a). *Experimentation and Evaluation Plans for the 2010 Census: Interim Report.* Panel on the Design of the 2010 Census Program of Evaluations and Experiments (CPEX). Committee on National Statistics. Washington, DC: The National Academies Press.

_______ (2008b). *Protecting Individual Privacy in the Struggle Against Terrorists—A Framework for Program Assessment.* Committee on Technical and Privacy Dimensions of Information for Terrorism Prevention and Other National Goals. Committee on Law and Justice, Committee on National Statistics, and Computer Science and Telecommunications Board. Washington, DC: The National Academies Press.

_______ (2008c). *Rebuilding the Research Capacity at HUD.* Committee to Evaluate the Research Plan of the Department of Housing and Urban Development, Center for Economic, Governance, and International Studies. Washington, DC: The National Academies Press.

_______ (2008d). *Surveying Victims: Options for Conducting the National Crime Victimization Survey.* Committee on National Statistics and Committee on Law and Justice. Washington, DC: The National Academies Press.

National Research Council and Institute of Medicine (1992). *Toward a National Health Care Survey: A Data System for the 21st Century.* Panel on the National Health Care Survey, Gooloo S. Wunderlich, ed. Committee on National Statistics and Division of Health Care Services. Washington, DC: National Academy Press.

_______ (2004). *Children's Health, The Nation's Wealth: Assessing and Improving Children's Health.* Committee on Evaluation of Children's Health and Board on Children, Youth, and Families. Washington, DC: The National Academies Press.

_______ (2008). *The National Children's Study Research Plan: A Review.* Panel to Review the National Children's Study Research Plan. Committee on National Statistics, Board on Children, Youth, and Families, and Board on Population Health and Public Health Practice. Washington, DC: The National Academies Press.

Norwood, Janet L. (1975). Should those who produce statistics analyze them? How far should analysis go? An American view. *Bulletin of the International Statistical Institute,* Proceedings of the 40th session, 46:420-432.

_______ (1995). *Organizing to Count: Change in the Federal Statistical System.* Washington, DC: The Urban Institute Press.

President's Commission on Federal Statistics (1971). *Federal Statistics. Vol. I.* Washington, DC: U.S. Government Printing Office.

Privacy Protection Study Commission (1977). *Personal Privacy in an Information Society.* Washington, DC: U.S. Government Printing Office.

Ryten, Jacob (1990). Statistical organization criteria for inter-country comparisons and their application to Canada. *Journal of Official Statistics* 6(3):319-332.

Triplett, Jack (1991). The federal statistical system's response to emerging data needs. *Journal of Economic and Social Measurement* 17(3, 4):155-201.

U.N. Economic Commission for Europe (2003). ECE-World Bank Seminar on the Application of the Fundamental Principles of Official Statistics. Almaty, Kazakhstan, April 28-29. Available: www.unece.org/stats/documents/ece-worldbank/2003/w1/1.e.pdf.

U.N. Statistical Commission (1994). *Fundamental Principles of Official Statistics.* Official Records of the Economic and Social Council, 1994, Supplement No. 9. New York: U.N. Statistical Commission. Available: www.unece.org/stats/documents/ece-worldbank/2003/w1/1.e.pdf.

_______ (2003). *Implementation of the Fundamental Principles of Official Statistics. Report of the Secretary-General.* New York: U.N. Statistical Commission. Available: www.unstats.un.org/unsd/dnss/gp/globreview.aspx.

U.N. Statistical Commission and Economic Commission for Europe (1991). Resolution on fundamental principles of official statistics in the Economic Commission for Europe (draft). *Report of the Thirty-Ninth Plenary Session,* June 17-21.

U.S. Census Bureau (1998). *Survey of Income and Program Participation Quality Profile 1998.* Third Edition. Prepared by Westat, Graham Kalton, project director. Washington, DC: U.S. Department of Commerce. Available: www.bls.census.gov/sipp/workpapr/230.pdf.

_______ (2003). *U.S. Census Bureau Data Stewardship/Privacy Impact Assessment—Demographic Surveys Program.* Washington, DC: U.S. Department of Commerce.

U.S. Departments of Agriculture, Commerce, Education, Energy, Health and Human Services, Justice, Labor, Transportation, and Treasury; National Science Foundation; and Social Security Administration (2002). Federal statistical organizations' guidelines for ensuring and maximizing the quality, objectivity, utility, and integrity of disseminated information. *Federal Register* 67(107):38467-38469.

U.S. General Accounting Office (1995). *Statistical Agencies: Adherence to Guidelines and Coordination of Budgets.* Washington, DC: U.S. Government Printing Office.

U.S. Office of Management and Budget (1985). Statistical policy directive on compilation, release, and evaluation of principle federal economic indicators. *Federal Register* 50(186):932-934. Available: www.whitehouse.gov/omb/inforeg/statpolicy/dir_3_fr_09251985.pdf.

_______ (1997). Order providing for the confidentiality of statistical information. *Federal Register* 62(124):35044-35050. Available: www.whitehouse.gov/omb/inforeg/conforder.pdf.

_______ (1999). *Statistical Programs of the United States Government, Fiscal Year 2000.* Statistical and Science Policy Office, Office of Information and Regulatory Affairs. Washington, DC: U.S. Government Printing Office.

_______ (2004a, December 16). *Issuance of OMB's "Final Information Quality Bulletin for Peer Review."* Memorandum from Joshua B. Bolten, Director, for heads of departments and agencies. Washington, DC. Available: www.whitehouse.gov/omb/memoranda/fy2005/m05-03.pdf.

_______ (2004b). *Statistical Programs of the United States Government, Fiscal Year 2005.* Statistical and Science Policy Office, Office of Information and Regulatory Affairs. Washington, DC: U.S. Government Printing Office.

_______ (2006a). *Guidance on Agency Survey and Statistical Information Collections.* Memorandum for the President's Management Council, John D. Graham, January 20. Available: www.whitehouse.gov/omb/inforeg/pmc_survey_guidance_2006.pdf.

_______ (2006b). Standards and Guidelines for Statistical Surveys, *Federal Register* 71(184):55522-55523. Complete document available: www.whitehouse.gov/omb/inforeg/statpolicy/standards_stat_surveys.pdf.

_______ (2007). Implementation guidance for Title V of the E-Government Act, Confidential Information Protection and Statistical Efficiency Act of 2002 (CIPSEA), *Federal Register* 72(115):33362-33377. Available: www.whitehouse.gov/omb/fedreg/2007/061507_cipsea_guidance.pdf.

_______ (2008a). *Budget of the United States Government, Fiscal Year 2009, Analytical Perspectives, Chapter 4: Strengthening Federal Statistics.* Washington, DC: U.S. Government Printing Office.

_______ (2008b). Statistical policy directive no. 4—Release and dissemination of statistical products produced by federal statistical agencies, *Federal Register* 73(46):12622-12626. Available: www.whitehouse.gov/omb/fedreg/2008/030708_directive-4.pdf.

_______ (2008c). *Statistical Programs of the United States Government, Fiscal Year 2009.* Statistical and Science Policy Office, Office of Information and Regulatory Affairs. Washington, DC: U.S. Government Printing Office.

Appendix A

Organization of the Federal Statistical System

This appendix provides a brief tour of the U.S. statistical system. It begins with an overview of the statistical system vis-à-vis the federal government as a whole. It then briefly summarizes the statistical functions of the U.S. Office of Management and Budget (OMB), the principal statistical agencies, and a selection of major statistical programs housed or sponsored by other agencies.

OVERVIEW

Budget

Levels and Trends

For fiscal year 2008 OMB estimates $5 billion in budget authority for government statistical programs, covering those carried out by designated statistical agencies and by policy, research, and program operation agencies, excluding the 2010 decennial census, which had another $1 billion in budget authority. This estimate includes the more than 100 agencies with direct funding for statistical activities of $500,000 or more, defined by OMB to include not only survey and census design and data collection, but also data analysis, forecasting, and modeling (U.S. Office of Management and Budget, 2008c:Table 1:3-4).[1]

Table A-1 shows approximately comparable estimates for 1998 and 2008 (in current and constant 2008 dollars) for statistical programs of all agencies with statistical activities of $500,000 or more and, separately, for the 14 member agencies of the Interagency Council on Statistical Policy (ICSP).[2] In 2008, the ICSP agencies accounted for 42 percent of the total budget authority for statistical activities, excluding the 2010 census; with the 2010 census authority included, they accounted for 52 percent.

The total budget authority for statistical activities increased in real terms by $905 million between 1998 and 2008, excluding the 2000 and 2010 censuses; with the censuses, the increase was $1.35 billion. Budget authority for the ICSP agencies, however, remained flat in real terms between 1998 and 2008, with some variation among agencies. In contrast, budget authority for the statistical activities of other agencies increased by 43 percent, so that the ICSP share of the total declined over the 10-year period from 50 to 42 percent without the censuses and from 56 to 52 percent with the censuses.

Comparisons of funding for the ICSP agencies for the two years (1998 and 2008) by department have to be made with caution because of the difficulty in ensuring comparability among the programs included in the "other" category in both years. For example, the Department of Health and Human Services (HHS) includes not only the National Center for Health Statistics, which experienced reduced funding over the period, but also equally large or larger agencies that conduct statistical programs, such as the Agency for Healthcare Research and Quality, which gained funding over the period. Many of the HHS agencies that reported statistical activi-

[1]"Direct funding" is directly appropriated to an agency. Some agencies (e.g., the Census Bureau) carry out statistical activities for other agencies on a cost-reimbursable basis. The funding for these activities is credited to the sponsoring agency and not to the data collection agency. OMB's annual compilation of statistical programs generally includes the entire budget for each of the 14 agencies represented on the ICSP; other agencies determine which parts of their budgets should be included according to the OMB definition of statistical activities.

[2]The nominal threshold of $500,000 for reporting statistical activities in 1998 is $734,500 in 2008 dollars, so that the budget authority for statistical activities may be underreported for 1998 compared with 2008. To the extent possible, the "other agencies" categories in Table A-1 are limited to agencies in existence in both years (exclusion of a few small agencies accounts for the difference between the $5,962.4 million reported total budget authority, including the census, for 2008 in Table A-1 and the $5,989.1 million reported by OMB).

TABLE A-1 Budget Authority, Statistical Programs of the U.S. Government, by Department, Fiscal Years 1998 and 2008 (millions of 2008 dollars)

Department or Agency	Fiscal Year 1998	Fiscal Year 2008	Percent Change[a]
Agriculture			
Economic Research Service*	103.0	77.4	–25
National Agricultural Statistics Service*	169.8	162.2	–4
Other agencies	238.2	255.7	+7
Commerce			
Bureau of Economic Analysis*	61.2	77.2	+26
U.S. Census Bureau, except decennial census*	450.7	463.1	+3
U.S. Census Bureau, decennial census*	555.3	1,004.1	+81
Other agencies	81.2	91.4	+13
Defense	11.9	15.8	+33
Education			
National Center for Education Statistics*	154.7	255.2	+65
Other agencies	N.A.	88.8	N.A.
Energy			
Energy Information Administration*	95.4	95.5	±0
Other agencies	34.5	16.3	–53
Health and Human Services			
National Center for Health Statistics*	121.7	113.6	–7
Other agencies	965.9	1,670.0	+73
Homeland Security	17.7	51.9	+193
Housing and Urban Development	42.7	50.6	+19
Interior	108.9	119.0	+9
Justice			
Bureau of Justice Statistics*	38.3	41.8	+9
Other agencies	19.0	22.5	+18
Labor			
Bureau of Labor Statistics*	546.8	544.3	±0
Other agencies	140.0	82.1	–41
Transportation			
Bureau of Transportation Statistics*	30.5	27.5	–1
Other agencies	99.0	112.4	+14
Treasury (Internal Revenue Service)			
Statistics of Income Division*	37.4	41.3	+10
Other agencies	17.0	N.A.	N.A.

Continued

TABLE A-1 Continued

Department or Agency	Fiscal Year 1998	Fiscal Year 2008	Percent Change[a]
Veterans Affairs	86.9	94.0	+8
Independent Agencies			
Office of Environmental Information, EPA*	207.1	118.1	–43
Office of Research, Evaluation and Statistics, SSA*	10.2	34.9	+242
Science Resources Statistics Division, NSF*	19.4	36.7	+89
Other agencies	143.2	199.0	+39
TOTAL, excluding decennial census	4,052.3	4,958.3	+22
Major statistical agencies (starred above)	2,046.2	2,088.8	+2
Other agencies	2,006.1	2,869.5	+43
TOTAL, including decennial census	4,607.6	5,962.4	+29
Major statistical agencies (starred above)	2,601.5	3,092.9	+19
Other agencies	2,006.1	2,869.5	+43

NOTES: Amounts represent actual and estimated direct budget authority for 1998 and 2008, respectively. 1998 dollars are converted into real 2008 dollars by the gross domestic product chain-type price indexes for federal government nondefense consumption expenditures of 93.4 in mid-1998 and 134.4 in mid-2008 (Table 3.10.4, line 34, at www.bea.gov/national/nipaweb/SelectTable.asp?Selected=N#S3). The amounts for the Environmental Protection Agency include all of EPA and not just the Office of Environmental Information. Funding for the National Center for Health Statistics Office of the Director is included in the 1998 amount but not the 2008 amount. Agencies that are not statistical agencies self-report to OMB on the activities they determine meet the OMB definition for reporting (see text). N.A. = No "other agencies" reported.
[a]Calculated as: (2008 – 1998) / 1998.
*Member, Interagency Council on Statistical Policy.
SOURCE: Data from U.S. Office of Management and Budget (1999:Table 1; 2008c:Table 1).

ties are part of the National Institutes of Health (NIH). Some NIH funding for statistical activities, such as for the surveys sponsored by the National Institute on Aging, is comparable to the activities of ICSP members, but much of the statistical work reported by NIH is not.

Not all of the work of ICSP agencies is carried out in-house. For fiscal year 2009, OMB estimates that 39 percent of the total budget authority of ICSP agencies will be used to purchase statistical services, such as data collection and analysis, from other organizations (U.S. Office of Management and Budget, 2008c:Table 3). About 28 percent of purchases (11 percent of total budget authority) will reimburse state and local governments for administrative records (e.g., birth and death records provided to the National

Center for Health Statistics and unemployment insurance wage records provided to the Bureau of Labor Statistics); almost 50 percent of purchases (19 percent of total budget authority) will be paid to private organizations for data collection and analysis services; and about 25 percent of purchases (9 percent of total budget authority) will be paid to other federal agencies, principally the Census Bureau. In dollar terms, the National Center for Education Statistics, the Bureau of Labor Statistics, and the National Center for Health Statistics dedicate the largest amounts of their budgets to purchasing statistical services; by percentage of budget authority, the National Center for Education Statistics, the Science Resources Statistics Division, the Bureau of Justice Statistics, and the National Center for Health Statistics are the largest users of purchased services. These patterns have remained roughly constant over the past decade (see U.S. Office of Management and Budget, 1999:Table 3).

Value

Spending on statistical programs is a tiny fraction of overall federal spending: In fiscal year 2008, the $5 billion in budget authority for all statistical programs identified by OMB amounted to 0.02 percent of the budget authority of about $2.5 trillion for the federal government. On a per capita basis, the $5 billion is equal to about $16 for every U.S. resident (305.7 million people in late 2008; see www.census.gov).

A basic public policy question is the value that the statistical system delivers for the federal government and the public. It is difficult to assign an overall valuation to the system or even to a specific agency or program (see National Research Council, 1985b:Ch. 3, App. 3A). A sense of value can be obtained in some instances by comparing the dollars spent on providing key statistics to the dollars that such statistics drive in the economy and society. For example, the Consumer Price Index (CPI) program of the Bureau of Labor Statistics had an estimated budget authority of $69 million in fiscal year 2008. Output from the CPI program is used for annual cost-of-living adjustments to payments for retirees and other beneficiaries under the Social Security program, which provided $585 billion in benefits to 50.4 million people in 2008 (one-sixth of the U.S. population; see http://www.ssa.gov): a difference of 1 percentage point in the CPI index amounts to about $6 billion in additional (or reduced) Social Security benefits in the subsequent year. Annual changes in the CPI also affect changes in commercial and residential rents, public and private sector wages, and components of the federal income tax code. Reports of monthly changes in the CPI are a major input for Federal Reserve Board decisions in setting short-term interest rates and to financial decisions throughout the public and private sectors. There are other such examples of consequential statistics throughout government and the economy.

Some statistical programs may lack clear-cut links to public and private sector financial outlays, but they nonetheless serve other important purposes:

- Providing information to inform policy makers and the public about the social and economic health of the nation, states, and localities—for example, the Bureau of Economic Analysis provides estimates of gross domestic product not only for the nation each quarter, but also for states and metropolitan areas each year, and the Census Bureau's American Community Survey provides estimates of educational attainment, median income, immigration, poverty, and many other characteristics for large and small geographic areas annually.
- Providing empirical evidence with which to develop and evaluate federal, state, local, and private-sector programs—for example, the American Housing Survey, sponsored by the Office of Policy Development and Research in the Department of Housing and Urban Development and conducted by the Census Bureau, provides valuable data on housing condition and housing finance with which to inform housing policy (see National Research Council, 2008c).
- Providing input to important social science research that, in turn, informs the public and policy making—for example, the National Long-Term Care Survey, funded by the National Institute on Aging, produced unexpected findings of declining disability rates for older Americans over time (see also National Research Council, 2005b).

Structure

The United States has a highly decentralized statistical system in comparison with other developed countries (see Norwood, 1995). Essentially, the system grew by adding separate agencies whenever the need for objective empirical information on a particular aspect of the economy, society, or environment came to the fore (see Part II). Periodic recommendations from presidential commissions and other initiatives to consolidate one or more of the principal statistical agencies have been ignored.

The statistical coordinating, clearance, review, and planning functions of the Statistical and Science Policy Office of OMB, which have their origins in the 1930s (see Appendix B), provide an important integrative force for the U.S. statistical system. Because, however, statistics on agriculture, education, health, justice, labor, and other topics are housed in agencies in different cabinet departments with different statutory provisions and are reviewed by different congressional committees, the system has limited capability to respond to changing priorities by such means as reallocating budgets across subject areas or to streamline agency operations by such means as sharing data (with some important exceptions in recently enacted legislation—see Appendix B).

Figure A-1 shows the major statistical programs in the executive branch of government by cabinet department: about 100 agencies or offices with statistical activities that have at least $500,000 direct budget authority in fiscal year 2008 (counting each NIH institute or center as a separate agency).

At the center of the system, in a sense, is the OMB Office of Information and Regulatory Affairs (OIRA), which includes the Statistical and Science Policy Office headed by the chief statistician of the United States, a senior executive civil service position. OIRA also includes the clearance officers who review individual survey and other information requests from most agencies; staff of the Statistical and Science Policy Office clear information requests from many of the principal statistical agencies and consult with the OIRA desk officers for the other agencies. Other parts of OMB recommend budgets for statistical agencies and programs in collaboration with the Statistical and Science Policy Office.

The chief statistician chairs the Interagency Council on Statistical Policy; the 14 agencies that are members of the ICSP are in nine cabinet departments and three independent agencies. Some of these agencies report directly to the secretary or other high-level official of their cabinet department; others are one, two, or even more layers further down the hierarchy; see Figure A-2. Several of these agencies have federal-state cooperative statistical programs that produce some of the nation's most important statistics, such as birth and death rates from vital records maintained by state registrars and estimates of employment from wage records maintained by state employment security offices.

The 14 agencies have their yearly budget requests reviewed and approved by seven different subcommittees of the House and Senate Appropriations Committees; see Figure A-2. The fact that different statistical

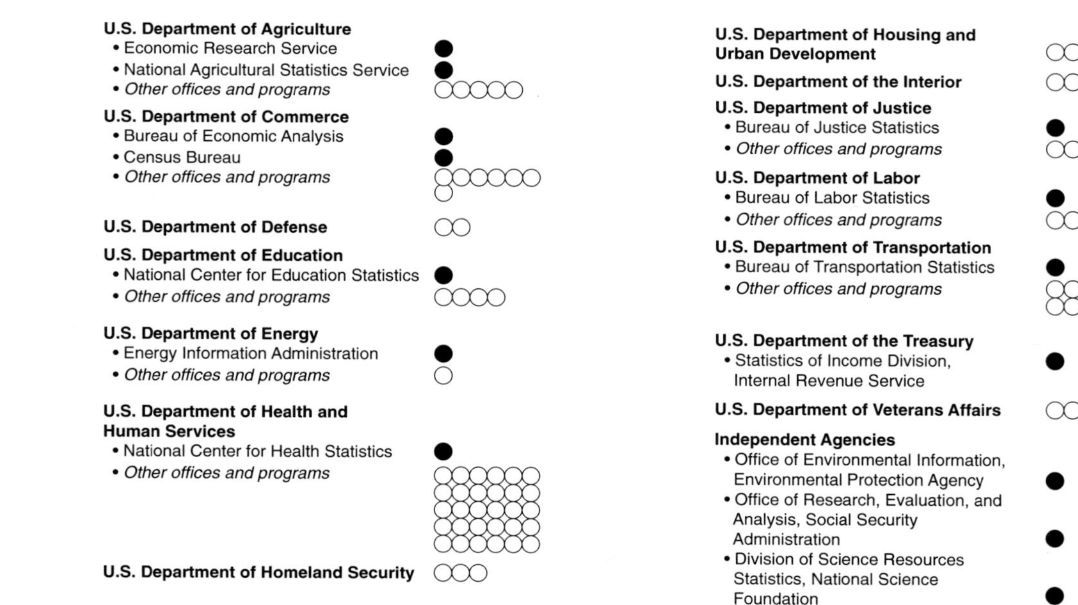

FIGURE A-1 Organization of principal federal statistical agencies and programs, by department, fiscal year 2008.
● Member, Interagency Council on Statistical Policy.
○ Offices or agencies with estimated spending on statistical activities of at least $500,000 in fiscal year 2008.
SOURCE: Based on U.S. Office of Management and Budget (2008c:Table 1).

agencies fall into different components of the federal budget for purposes of annual congressional appropriations complicates the possibility of coordination of statistical programs across the government.

Finally, there are some important federal agencies that have statistical activities that are not included in the OMB annual compilation because they are not part of the executive branch. These agencies include the Congressional Budget Office, which develops and applies projection models for the budgetary impact of current and proposed federal programs; the Federal Reserve Board, which compiles the widely used Flow of Funds Report and other statistical series and periodically conducts the Survey of Consumer Finances; and the U.S. Government Accountability Office, which uses statistical data in evaluations of government programs.

U.S. OFFICE OF MANAGEMENT AND BUDGET

The 1995 reauthorization of the Paperwork Reduction Act of 1980 and prior legislation give the Office of Information and Regulatory Affairs the authority to approve all agency information collection requests, including all survey and other statistical information requests. OIRA also reviews all proposed federal regulations. The chief statistician's office in OIRA (the Statistical and Science Policy Office) establishes statistical policies and standards, identifies priorities for improving programs, evaluates statistical programs for compliance with OMB guidance, reviews statistical agency budgets, approves information collections for many of the principal statistical agencies, provides guidance to OIRA desk officers who review statistical information requests from other federal agencies, and coordinates U.S. participation in international statistical activities.

As required by the Paperwork Reduction Act, the office annually puts out *Statistical Programs of the United States Government* (the "blue book;" see U.S. Office of Management and Budget, 2008c). It also prepares a chapter each year in the *Analytical Perspectives* volume of the President's budget, which provides a cross-cutting analysis of the budget requests for the principal statistical agencies (see U.S. Office of Management and Budget, 2008a). The chief statistician's office currently has a staff of six professionals, some of whom focus largely on science policy.

The chief statistician chairs the ICSP, which began operating informally in the late 1980s and was authorized in statute in the 1995 reauthorization of the Paperwork Reduction Act. The chief statistician's office also sponsors the Federal Committee on Statistical Methodology and other bodies that

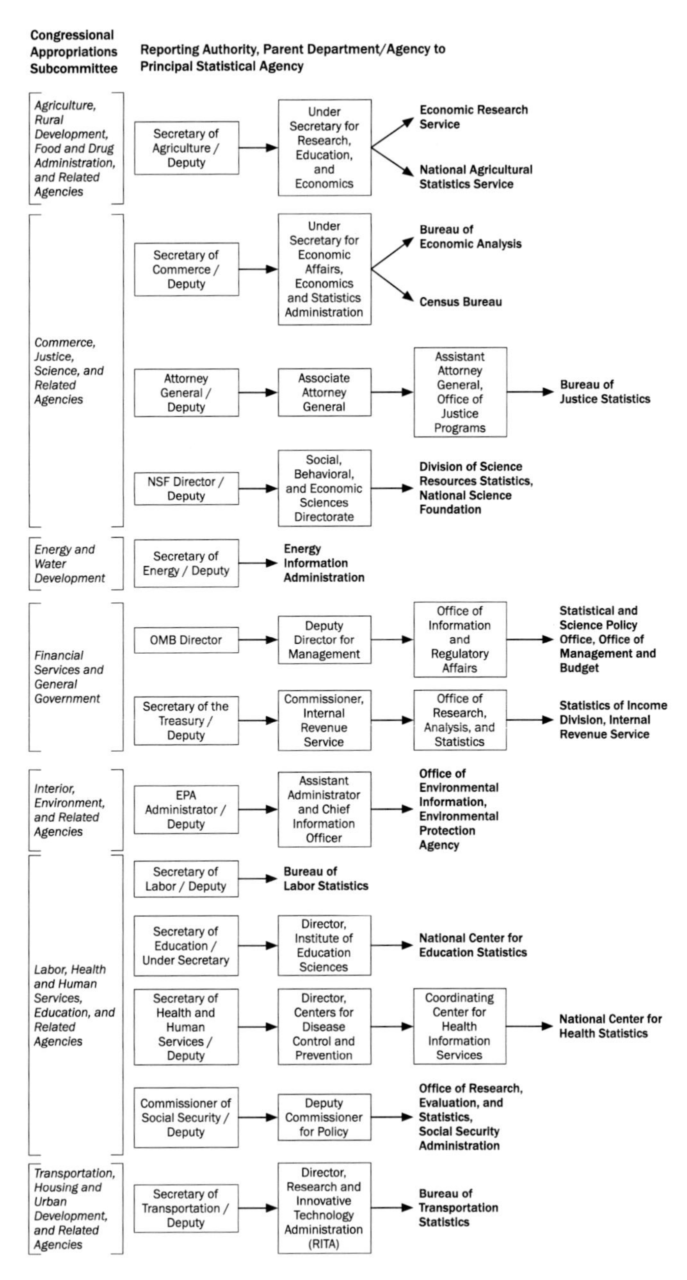

FIGURE A-2 Members of the Interagency Council on Statistical Policy: Organizational Location and Relevant Congressional Appropriations Subcommittee based on subcommittee jurisdictions in the 110th Congress.

help to coordinate the statistical system, such as the Interagency Forum on Aging-Related Statistics and the Interagency Forum on Child and Family Statistics (see discussion in Part II under Practice 11).

Appendix B provides background information on the Paperwork Reduction Act, statistical policy directives issued by the chief statistician's office, and other legislation that the office oversees for the U.S. statistical system.

PRINCIPAL STATISTICAL AGENCIES

This section provides information—primarily from agency web sites (see Appendix D) and OMB publications—on 13 of the 14 members of the ICSP, excluding only the Office of Environmental Information in the Environmental Protection Agency, which is not a self-contained statistical unit. The information provided for the 13 agencies includes origins, authorizing legislation or other authority, status of head (presidential appointee, career senior executive service official), full-time permanent staffing levels (see U.S. Office of Management and Budget, 1999, 2008c:App. B), and principal programs (see Table A-1 for budget levels and trends). The agencies are discussed in alphabetical order.

Bureau of Economic Analysis

The Bureau of Economic Analysis (BEA), along with the Census Bureau, is part of the Economics and Statistics Administration in the Department of Commerce, headed by the Under Secretary for Economic Affairs. BEA's director is a career senior executive service appointee; the bureau has a full-time staff of about 500 people (compared with about 420 people in 1998).

BEA's history traces back to 1820 when the Secretary of the Treasury was directed by Congress to compile and publish statistics on U.S. foreign commerce. Three 20th-century predecessors of BEA were all located in the Department of Commerce: the Bureau of Statistics (1903-1912); the Bureau of Foreign and Domestic Commerce (1912-1945); and the Office of Business Economics (1945-1972).

BEA produces statistics on the performance of the nation's economy. Although BEA collects some source data, it primarily compiles data from the Census Bureau, the Bureau of Labor Statistics, and other agencies as input to estimating the National Income and Product Accounts (NIPAs), which include estimates of the nation's gross domestic product (GDP) and related measures. The GDP, which was recognized by the Department of Commerce as its greatest achievement of the 20th century, is highly influential on U.S. financial markets.

Since the NIPAs were first developed in the aftermath of the Great Depression, BEA has extended its estimates to cover a wide range of economic activities for the nation, regions, and industries and also for the nation's position in the world economy. BEA also conducts research and development on "satellite accounts" in such areas as health care, transportation, and research and development investments. Satellite accounts enable experimentation with new accounts before they are ready to incorporate into the main accounts and with nonmarket sectors that are not part of the market-based NIPAs.

Bureau of Justice Statistics

The Bureau of Justice Statistics (BJS) was formally established in December 1979 by the Justice Systems Improvement Act of 1979 (P.L. 96-157), inheriting statistical functions that had previously been vested in an office of the Law Enforcement Assistance Administration. Administratively, BJS is housed in the U.S. Department of Justice's Office of Justice Programs (OJP), which also contains the National Institute of Justice (a research agency) and other agencies that are primarily focused on providing grant and technical assistance to state and local governments and law enforcement agencies. BJS's director is a presidential appointee and reports to an assistant attorney general for OJP. The bureau has a full-time staff of about 55 (about the same number as it had in 1998).

The centerpiece of BJS's data collections is the National Crime Victimization Survey (NCVS; originally the National Crime Survey), which has served as one of the nation's principal measures of crime (particularly crime not reported to police) since its full-scale implementation in 1972. Data collection for most BJS surveys is conducted by the Census Bureau or private contractors. OMB estimates that 86 percent of BJS's 2009 budget authority will be spent on purchased services (the same percentage as a decade earlier). BJS publishes annual statistics on criminal victimization, populations under correctional supervision, law enforcement management and administration, and case processing in the state and federal courts. Its periodic data series cover the administration of law enforcement agencies and correctional facilities, prosecutorial practices and policies, state court case processing, felony convictions, criminal justice expenditure and employment, civil case processing in state courts, and special studies on other criminal justice topics.

Bureau of Labor Statistics

The Bureau of Labor Statistics (BLS) is an agency of the Department of Labor. It is responsible for the production of some of the nation's most sensitive and important economic data, which serves the public, Congress, other federal agencies, state and local governments, business, and labor. The BLS commissioner is a presidential appointee with a fixed term of 4 years; the agency has a staff of about 2,400 people, down from about 2,600 people in 1998.

The history of the BLS dates back to 1884, when the Bureau of Labor was established in the Interior Department to collect information about employment and labor. It was made an independent (subcabinet) agency by the Department of Labor Act in 1888; it was made part of the Department of Commerce and Labor (as the Bureau of Labor) in 1903 and transferred to the newly created Department of Labor in 1913.

BLS programs use a variety of data collection methods and sources. Certain wage, benefit, employment, and price data are collected by BLS staff located throughout the country, who contact employers, households, and businesses directly. BLS has contractual arrangements with various state agencies to collect much of its employment and workplace safety and health data. Contractual arrangements with the Census Bureau support collection of several programs, including the Current Population Survey (the source of monthly unemployment statistics) and the Consumer Expenditure Survey (the source of market baskets for the Consumer Price Index). Some BLS data, such as data for the National Longitudinal Survey, are collected by private contractors. Finally, certain BLS data, such as information on work stoppages, rely on information from secondary sources.

BLS's surveys, indexes, and statistics fall into four main categories:

(1) Consumer expenditures and prices, including the CPI, producer price index (PPI), U.S. import and export prices indexes, as well as consumer expenditures.

(2) The labor force, including monthly data on employment from households and business establishments, monthly and periodic data on unemployment, time use, job openings and labor turnover, occupational

employment, and mass layoffs, and longitudinal data on the work experience of cohorts of the population.

(3) Compensation and working conditions, including the employment cost index, and workplace injuries and fatalities.

(4) Productivity.

BLS also projects trends in occupational employment.

Bureau of Transportation Statistics

The Bureau of Transportation Statistics (BTS) is an agency in the Research and Innovative Technology Administration (RITA) of the Department of Transportation. RITA also includes the Intelligent Transportation Systems Joint Program Office; the Office of Research, Development, and Technology; the Transportation Safety Institute; and the Volpe National Transportation Systems Center.

BTS's director is now a career senior executive service appointee who reports to the head of RITA; prior to 2004, the director was a presidential appointee with a fixed term of 4 years who reported directly to the Secretary of Transportation. BTS has a staff of about 120 people (an increase from about 50 people in 1998).

BTS was established by the 1991 Intermodal Surface Transportation Efficiency Act (ISTEA) and began operations in late 1992. It was moved to the newly created RITA by the Norman Y. Mineta Research and Special Programs Improvement Act of 2004. Prior to the establishment of BTS, statistical programs of the Department of Transportation focused exclusively on specific modes of transportation (highways, airlines, railroads, etc.), except for the first 10 years of the department's existence (1967-1977), when the Office of the Secretary funded intermodal surveys on commodity flows and long-distance personal transportation.

BTS is charged to produce an annual report on transportation statistics, develop intermodal data on commodity and passenger flows, administer the National Transportation Library, and carry out other functions to ensure that the department, the states, and other federal agencies have available comprehensive information on the nation's transportation systems. BTS also operates the Office of Airline Information, which was transferred to it from the now-defunct Civil Aeronautics Board. BTS contracts with the Census Bureau for major surveys.

Census Bureau

The Census Bureau, along with BEA, is part of the Economics and Statistics Administration in the Department of Commerce, headed by the Under Secretary for Economic Affairs. It conducts population and economic censuses and a wide array of surveys.

Population censuses are required by the U.S. Constitution to be conducted every 10 years for reapportioning seats among the states in the House of Representatives. The first censuses were conducted by U.S. marshals under the authority of the Secretary of State. Beginning in 1850, a separate census office was established each decade to supervise the census. In 1902 a permanent Census Bureau was established; it was made part of the new Department of Commerce and Labor in 1903 and moved to the newly created Department of Commerce in 1913. Title 13 of the U.S. Code codifies all of the legislation that pertains to the Census Bureau, including strict provisions for protecting the confidentiality of population and business information.

The director of the Census Bureau is appointed by the president. The bureau has about 4,800 staff, excluding staff hired for the decennial censuses (an increase from about 3,800 staff in 1998).

The major periodic activity of the Census Bureau is the decennial population census, which in 2010 will consist of basic questions on age, sex, race, Hispanic origin, relationship to household head, and housing tenure (own, rent). The Bureau also conducts the continuous American Community Survey (ACS), which replaced the content included for a sample of households in the census from 1940 through 2000 (the "long form"), a large number of household surveys (most under contract for other agencies), and a panoply of censuses and surveys of business establishments and governments. The Bureau produces annual population and housing estimates (in cooperation with state and local governments), estimates of poverty, median income, and health insurance coverage using statistical models for small areas, and geographic products based on its Topologically Integrated Geographic Encoding and Referencing (TIGER) system.

The Census Bureau has a substantial portfolio of reimbursable work for other agencies, which is primarily for the conduct of surveys, such as the Current Population Survey, the American Housing Survey, the Consumer Expenditure Survey, and many others. OMB estimates that reimbursable work will account for almost $250 million of the Bureau's fiscal year 2009 budget, which represents more than 50 percent of its budget authority excluding the 2010 census and ACS. The amount of reimbursable work for the Census Bureau was about the same in real terms in 1998 (see U.S. Office of Management and Budget, 1999:Table 2, 2008c).

Economic Research Service

The Economic Research Service (ERS), along with the National Agricultural Statistics Service, reports to the Under Secretary for Research, Education, and Economics in the U.S. Department of Agriculture (USDA). The administrator of ERS is a career senior executive service appointee; the agency has a staff of about 430 (a decrease from about 510 in 1998).

The origins of ERS trace back to 1905, when USDA established the Office of Farm Management, renamed the Office of Farm Management and Farm Economics in 1919. The Office's research areas included farm organization, cost of production, farm labor, farm finance, land economics, agricultural history, and rural life studies. In 1922, USDA established the Bureau of Agricultural Economics (BAE), which not only conducted research, but also, in the 1930s, served as the central planning agency for department policy.

In 1953 USDA centralized agricultural policy planning in its administrative office and reassigned the economic research and service functions of BAE to two new agencies, the Agricultural Marketing Service and the Agricultural Research Service. In 1961, USDA created the Economic Research Service (ERS) to concentrate economic research in a single agency.

Today, ERS conducts economic research and policy analysis that informs program and policy decisions throughout USDA. The agency's mission is to anticipate food, agricultural, agri-environmental, and rural development issues that are on the horizon and conduct peer-reviewed economic research so that research findings are available when issues require decisions by policy makers. ERS does not make recommendations; instead, it designs its research to demonstrate to users the consequences of taking alternative policy or programmatic pathways.

ERS is also the primary source of statistical indicators that, among other things, gauge the health of the farm sector (including farm income estimates and projections), assess the current and expected performance of the agricultural sector (including trade), and provide measures of food insecurity in the United States and abroad. ERS jointly funds with the National Agricultural Statistics Service a major survey on farm household income and crop practices, the Agricultural Resources Management Survey (ARMS).

Energy Information Administration

The Energy Information Administration (EIA) is an agency of the Department of Energy (DOE); its administrator is a presidential appointee. EIA has a staff of about 360 (about the same number as in 1998).

EIA was created by Congress in 1977 as part of the newly established Department of Energy. Its mission is to provide policy-independent energy data, forecasts, and analyses in order to promote sound policy making, efficient markets, and public understanding regarding energy and its interaction with the economy and the environment. To assure EIA's independence, the Department of Energy Organization Act specifies that EIA's products are not subject to clearance by executive branch officials; in particular, the administrator does not need to obtain the approval of any other DOE official for data collection and analysis, and he or she does not need to obtain the approval of anyone in DOE or elsewhere in the executive branch before publishing energy data and analysis reports.

Many EIA data products, such as weekly, monthly, and annual data on petroleum and natural gas supply, deal with specific industries; others contain data on all fuel types. Participation in EIA energy surveys is mandatory. EIA conducts surveys, using private contractors, of energy producers, users, and transporters, and certain other businesses. Data on energy consumption are collected for households, commercial buildings, manufacturing, and transportation. Analyses prepared by EIA staff cover energy economics, technology, production, prices, distribution, storage, consumption, and environmental effects. EIA forecasts cover all energy types and include supply, consumption, prices, and other factors; short-term forecasts go out 6-8 quarters into the future; 20-year projections are also developed and often serve as the baseline for independent analyses of policy proposals that are prepared by EIA at the request of Congress or the administration. More than three-quarters of EIA's resources are used for energy data collection and dissemination; the rest are used to support forward-looking forecasts, projections, and analyses.

National Agricultural Statistics Service

The National Agricultural Statistics Service (NASS), along with ERS, reports to the Under Secretary for Research, Education, and Economics in USDA. The administrator of NASS is a career senior executive service appointee; NASS has about 1,100 staff (about the same number as in 1998).

USDA, which President Lincoln called "the people's department," was established in 1862; in 1863, it established a Division of Statistics, which began publishing monthly crop reports. A scandal in 1905 regarding leaks of cotton crop forecasts to a financier led to legislation in 1909 that made premature disclosure of agricultural statistical reports a criminal offense and established the Crop Reporting Board (now the Agricultural Statistics Board) to oversee the physical security of the reports and to provide expert judgment on crop forecasts. A USDA reorganization in 1961 led to the creation of the Statistical Reporting Service, known today as the National Agricultural Statistics Service (NASS), of which the Agricultural Statistics Board is a part.

NASS acquired responsibility for the Census of Agriculture beginning in 1997, collecting the data through a contract with the Census Bureau for handling the mail questionnaires and by using its field office staff to perform detailed editing, follow-up, and analysis. Previously, the agriculture censuses were the responsibility of the Census Bureau with extensive input from NASS. The agriculture census dates back to 1840, when it was conducted as part of the decennial population census; the agriculture census was made quinquennial beginning in 1925.

NASS works with its field offices to carry out hundreds of surveys every year and prepares reports covering virtually every aspect of U.S. agriculture. Examples include production and supplies of food and fiber, prices paid and received by farmers, farm labor and wages, farm finances, chemical use, and changes in the demographics of U.S. producers. Interviewing staff are obtained through contracting with the National Association of State Departments of Agriculture.

National Center for Education Statistics

The National Center for Education Statistics (NCES), along with three research and evaluation centers, is part of the Institute of Education Sciences in the Department of Education. Its commissioner is a presidential appointee for a fixed term of 6 years; it has a staff of about 110 people (about the same number as in 1998).

NCES's origins date back to 1867 when Congress established a Department of Education and gave it a primary mission of "collecting such statistics and facts as shall show the condition and progress of education in the several States and Territories, and of diffusing such information respecting the organization and management of schools and school systems and methods of teaching." The legislation also charged the Commissioner of Education to issue an annual report. However, only 2 years later the department was abolished and the Office of Education was transferred to the Department of the Interior, where it remained through 1939. The Office of Education was part of the newly created Federal Security Agency from 1939 to 1953, when it was made part of the newly created Department of Health, Education, and Welfare. A separate Department of Education was established in 1980.

A major function of the Office of Education throughout its history has been the collection and publication of education statistics. NCES was established in 1965 as a staff office reporting to the Commissioner of Education. NCES received statutory authority in 1974; in 1980 it was made part of the Office of Educational Research and Improvement, which in 2002 became the Institute of Education Sciences. Supporting the independence of NCES, the Education Sciences Reform Act of 2002, which created IES (20 USC §9517(d)), stipulated that "each Commissioner, except the Commissioner for Education Statistics, shall carry out such Commissioner's duties . . . under the supervision and subject to the approval of the Director of IES."

NCES has an extensive survey program, including longitudinal surveys that follow the educational experience of cohorts of the U.S. population from early childhood through adulthood, periodic surveys of adult literacy, and international studies of educational achievement. It also collects the "Common Core of Data" from administrative records of state and local K-12 educational agencies and the Integrated Postsecondary Education Data System. It regularly assesses the educational knowledge and achievement of primary and secondary school students in the National Assessment of Educational Progress (NAEP). It also administers the Statewide Data Systems Program, which provides grants to the states to develop longitudinal databases of student records for analyzing student performance and for identifying methods to improve achievement.

NCES contracts for a substantial portion of its work, including not only data collection, but also data analysis and preparation of reports. In 2009, 97 percent of its estimated budget authority will be used for data and analysis from state agencies, the Census Bureau, and private contractors (about the same percentage as a decade ago).

National Center for Health Statistics

The National Center for Health Statistics (NCHS) is part of the Centers for Disease Control and Prevention (CDC) in HHS. The director of NCHS is a career senior executive service appointee; NCHS has a staff of about 520 people (about the same number as in 1998).

NCHS's roots lie in two formerly separate historical strands of the provision of national health statistics. The first strand is vital statistics on births, deaths, and other life events; it traces back to 1902, when Congress gave the newly created permanent Census Bureau the authority to establish registration areas to produce nationally comparable vital statistics by working with state agencies. This function was transferred in 1946 to the Federal Security Administration, which was folded into the new Department of Health, Education, and Welfare in 1953.

The second strand is general statistics on the nation's health, which were provided for in the 1956 National Health Survey Act. NCHS was created in 1960 as the merger of the National Office of Vital Statistics and the National Health Survey Division; it was relocated every few years in HHS until its last relocation in 1987, when it was made part of CDC. In 2005 it became one of three centers reporting to the newly created Coordinating Center for Health Information and Service in CDC.

NCHS has four major programs:

(1) The National Health Interview Survey (NHIS), in continuous operation since 1956, which collects a wide range of information on self-reported health status and conditions and use of healthcare services by the population.

(2) Surveys of healthcare providers, including nursing homes, hospitals, and outpatient facilities.

(3) The National Health and Nutrition Examination Survey (NHANES), which not only ascertains self-reported information on health and dietary intake, but also, by use of mobile examining units, obtains extensive information from physical examinations and laboratory tests.

(4) Vital statistics.

In 2009, according to OMB, 76 percent of NCHS's estimated budget authority will be used to purchase data collection and reporting services from state and local governments, the Census Bureau, and private contractors (slightly more than a decade earlier).

Office of Research, Evaluation, and Statistics

The Office of Research, Evaluation, and Statistics (ORES), along with eight other offices, reports to the Deputy Commissioner for Retirement and Disability Policy of the Social Security Administration (SSA), which became independent from HHS in 1995. ORES is headed by an associate commissioner, who is a career senior executive service appointee; it has a staff of about 100 people (comparable information is not available for 1998).

SSA began as the Social Security Board in 1935; it became part of the Federal Security Agency in 1939, part of the Department of Health, Education, and Welfare in 1953, and part of HHS in 1980; it regained independent agency status in 1995. From the outset, SSA has had a research, statistics, and evaluation function.

ORES conducts research and evaluation on the effects of Social Security and Supplemental Security Income and proposed changes in those programs on individuals, the economy, and program solvency. It provides data on program benefits and covered workers and develops and operates microsimulation models that estimate the costs and distributional effects of proposed changes in Social Security programs. Periodically, it has sponsored surveys of new beneficiaries, linked with SSA administrative records.

Science Resources Statistics Division

The Science Resources Statistics Division (SRS) is part of the Social, Behavioral, and Economic Sciences Directorate of the National Science Foundation (NSF). Its director is a career senior executive service appointee; it has a staff of about 50 people (comparable information is not available for 1998).

At its founding in 1950, NSF was charged to maintain a register of scientific and technical personnel so that the nation would be able to mobilize the scientific and technical work force in the event of a major war. Although no longer required to maintain a complete register, NSF has continued (by the terms of the National Science Foundation Act of 1950, as amended) to have responsibility "to provide a central clearinghouse for the collection, interpretation, and analysis of data on scientific and engineering resources and to provide a source of information for policy formulation by other agencies of the Federal Government." NSF also has a congressional mandate from 1980 to provide information on women and minorities in science and engineering.

The NSF mandates provide the basis for two major statistical programs in SRS: censuses and surveys of people trained in or working in science and engineering positions and of new bachelor's graduates, new master's graduates, and doctoral recipients in science and engineering fields; and surveys of research and development (R&D) expenditures by private industry and academic institutions and of R&D funding by the federal government. To support its programs, 87 percent of SRS's estimated budget authority in 2009 will be used to purchase data collection and other services from the Census Bureau and private contractors (comparable information for a decade ago is not available). SRS also serves as staff to the National Science Board and produces the biennial congressionally mandated *Science and Engineering Indicators Report*, which uses data from all 11 of the SRS surveys.

Statistics of Income Division

The Statistics of Income Division (SOI), along with four other units, is part of the Office of Research, Analysis, and Statistics of the Internal Revenue Service (IRS) in the Department of the Treasury. The director is a career senior executive service appointee; it has a staff of about 170 people and funds an additional 250 IRS field positions for statistical processing purposes (comparable information is not available for 1998).

SOI's history traces back to the enactment of authority to levy individual income taxes in the 16th amendment to the U.S. Constitution, which was ratified in 1913. The Revenue Act of 1916 mandated the annual publication of statistics related to the "operations of the internal revenue laws."

SOI provides annual income, financial, and tax data, based largely on individual and corporate tax returns and on returns filed by most tax-exempt organizations. It also provides periodic data derived from other returns and schedules, such as estate and gift tax returns and schedules of gains and losses from sales of capital assets. SOI data are available to staff in the Department of the Treasury and the Congressional Joint Committee on Taxation for policy analysis and revenue estimation, and to the Congressional Budget office for modeling Social Security and Medicare programs. Selected tax return data are also available, under strict confidentiality protection provisions, for use by the Census Bureau, the Bureau of Economic Analysis, and the National Agricultural Statistics Service in conducting business and agriculture censuses and surveys and producing the National

Income and Product Accounts. (See discussion in Appendix B of the Confidential Information Protection and Statistical Efficiency Act.)

OTHER MAJOR STATISTICAL PROGRAMS

This section lists and briefly describes six major statistical programs that are conducted or sponsored by agencies of the federal government other than the principal statistical agencies. The intent is to illustrate the range of the federal government's statistical portfolio.

- **Medical Expenditure Panel Survey (MEPS)**—MEPS is a statistical program of the Agency for Healthcare Research and Quality (AHRQ) in the Department of Health and Human Services. MEPS is the core healthcare expenditure survey in the United States with a primary analytical focus directed to the topics of healthcare access, cost, and coverage. MEPS was designed to provide data for healthcare policy analysis and research; it was first conducted in 1977 and again in 1987 (under different names) and became a continuous survey in 1996. MEPS consists of a family of three interrelated surveys: the Household Component (HC), the Medical Provider Component (MPC), and the Insurance Component (IC). The household survey collects information from household members and their health care providers and employers in order to construct a complete picture of medical care use, expenditures, and health insurance coverage and reimbursements. Households are in a MEPS panel for five rounds of interviewing that cover 2 years, so that patterns of medical care and expenditures can be observed over time; a new household panel begins every year. Data for the MEPS household and medical provider surveys are collected by private contractors; the household survey sample of about 14,000 households per year is drawn from the NCHS National Health Interview Survey. In addition, the MEPS Insurance Component (IC) collects data each year from a sample of about 30,000 private and public sector employers on the health insurance plans they offer their employees. The collected data include the number and types of private insurance plans offered (if any), premiums, contributions by employers and employees, eligibility requirements, benefits associated with these plans, and employer characteristics. Data for the MEPS IC are collected by the Census Bureau. The MEPS budget accounted for about one-sixth of the total AHRQ budget of about $330 million in 2008.

- **National Agricultural Workers Survey (NAWS)**—NAWS is a survey of the Employment and Training Administration (ETA) in the Department of Labor, which provides data on wage and migration history, type of crops worked, unemployment benefits, housing, health care, use of public programs, and other characteristics of the U.S. crop labor force. The information, which is used by numerous federal agencies for occupational injury and health surveillance, estimating the need for services for workers, allocating program dollars to areas of greatest need, and program design and evaluation, is obtained directly from farm workers through personal interviews. Since 1988, when the survey began, nearly 50,000 workers have been interviewed. The survey samples crop workers in three cycles each year to reflect the seasonality of agricultural production and employment. Workers are located at their farm job sites. During the initial contact, arrangements are made to interview the respondent at home or at another location convenient to the respondent. Depending on the information needs and resources of the various federal agencies that use NAWS data, between 1,500 and 4,000 workers are interviewed each year. NAWS is an important although small component of the ETA budget of $9 billion, 95 percent of which reimburses state agencies for data from administrative records on employment and wages.

- **National Automotive Sampling System (NASS)**—NASS is an administrative-records-based data collection system of the National Center for Statistics and Analysis (NCSA) of the National Highway Traffic Safety Administration (NHTSA) in the Department of Transportation. NASS was created in 1979 as part of a nationwide effort to reduce motor vehicle crashes, injuries, and deaths on U.S. highways. NASS samples accident reports of police agencies within randomly selected areas of the country. For the Crashworthiness Data System component of NASS, NCSA field researchers collect detailed information from police accident reports for selected crashes on exterior and interior vehicle damage, occupant injury, environmental conditions at the time of the crash, and other characteristics. For the General Estimates System component of NASS, only basic information is recorded from the police accident reports for a larger sample of crashes. The NASS infrastructure is also used for special studies and surveys, such as a recently completed National Motor Vehicle Crash Causation Survey (NMVCCS), which sampled police accident reports in real time and obtained on-scene information in addition to the information reported by the police.

NHTSA's overall budget is about $840 million; the budget for the National Center for Statistics and Analysis was about $36 million in 2008 and will decrease to about $28 million in 2009 with the completion of the NMVCCS. The National Automotive Sampling System's budget is about $12 million.

- **National Resources Inventory (NRI)**—The NRI is a statistical program of the Natural Resources Conservation Service (NRCS) in the Department of Agriculture. The current NRI is a longitudinal survey of soil, water, and related environmental resources designed to assess conditions and trends on non-federal U.S. land parcels. NRCS has conducted the NRI in cooperation with the Iowa State University Center for Survey Statistics and Methodology since 1977. The NRI was conducted on a 5-year cycle during the period 1982 to 1997 but beginning in 2000 is now conducted annually. NRI data were collected every 5 years for 800,000 sample sites; annual NRI data collection occurs at slightly less than 25 percent of these same sample sites. Data collected on conditions for the same sites year by year enable analysis of the effects of resource conservation programs and other applications. The NRI is a small but valuable component of NRCS's budget of about $800 million.

- **National Survey on Drug Use & Health (NSDUH)**—NSDUH is a continuing survey of the Office of Applied Studies (OAS) in the Substance Abuse and Mental Health Services Administration (SAMHSA). It is the nation's primary data system for collecting data on the incidence and prevalence of substance abuse and adverse health consequences associated with drug abuse from the civilian, noninstitutionalized population of the United States ages 12 and older. NSDUH (formerly called the National Household Survey on Drug Abuse) was fielded periodically from 1972 to 1990 and annually beginning in 1991. The completed sample size each year is about 67,500 people, with oversampling of teenagers and young adults; data collection is by a private contractor. The budget for statistical activities of SAMHSA, which includes not only NSDUH, but also the Drug and Alcohol Services Information System and its associated surveys (the primary data source for information on the nation's treatment system and outcomes), the Drug Abuse Warning Network (DAWN) (a public health surveillance system, which monitors drug-related visits to hospital emergency departments, as well as drug-related deaths investigated by medical examiners and coroners), and other programs, is about $130 billion of a total SAMHSA budget of about $3.2 billion.

- **Panel Study of Income Dynamics (PSID)**—The PSID is a longitudinal survey that began in 1968 and has followed several thousand families ever since that time. It is conducted by the Survey Research Center of the Institute for Social Research of the University of Michigan. While the PSID's original funding agency was the Office of Economic Opportunity of the Department of Commerce, the study's major funding source is now the National Science Foundation. Substantial additional funding has been provided by the National Institute on Aging, the National Institute of Child Health and Human Development, the Office of the Assistant Secretary for Planning and Evaluation of the Department of Health and Human Services, the Economic Research Service of the Department of Agriculture, the Department of Housing and Urban Development, the Department of Labor, and the Center on Philanthropy at the Indiana University-Purdue University, Indianapolis.

The PSID emphasizes the dynamic aspects of economic and demographic behavior, but its content is broad, including sociological and psychological measures. From 1968 to 1996, the PSID interviewed individuals in the original sample of about 4,800 families every year, whether or not they were living in the same dwelling or with the same people. In 1997 interviewing was changed to every other year, the original sample was reduced, and a sample of Hispanic families added in 1990 was replaced by a sample of post 1968 immigrant families and their adult children of all ethnic groups. The current sample of families, including those formed by children leaving their parental home, is about 8,000.

Citation studies show that since 1968, more than 2,000 journal articles, books and book chapters, government reports, working papers, and dissertations have been based upon the PSID. The PSID was founded to study poverty and the effects of programs to combat poverty, and an important early finding was that family structure changes such as divorce are as important to family well-being as employment. As the survey has added content and extended its period of observation, the data have also contributed importantly to studies of intergenerational patterns of work, welfare receipt, and other behaviors, international comparisons with panel data from other countries, neighborhood effects on family well-being (using data files augmented with census-based characteristics of sample members' communities), and long-term trends in marital and fertility histories and living arrangements.

Appendix B

Legislation and Regulations That Govern Federal Statistics

This appendix summarizes the major legislation and agency regulations and guidance that govern the operations of the federal statistical system as a whole:

- The 1980 Paperwork Reduction Act, as reauthorized in 1995, and associated guidance (the discussion also covers the 1942 Federal Reports Act and the 1950 Budget and Accounting Procedures Act).
- The 2000 Information Quality Act and associated guidelines.
- The 1997 Office of Management and Budget (OMB) Order Providing for the Confidentiality of Statistical Information.
- The 2002 Confidential Information Protection and Statistical Efficiency Act (CIPSEA—Title V of the E-Government Act) and associated guidance.
- The 2002 E-Government Act, Section 208, which requires privacy impact assessments for federal data collections and associated guidance (the discussion also covers the 1974 Privacy Act).
- The 2002 Federal Information Security Management Act (FISMA—Title III of the E-Government Act).
- 2004 OMB peer review guidance.
- 2002-2008 OMB provisions for rating the performance of federal programs using the Program Assessment Rating Tool (PART).
- OMB statistical policy directives.

Most of this legislation, regulations, and guidance pertains to the authority of OMB, which plays a critical role in oversight of the federal government's widely dispersed statistical operations. The oversight dates to 1939, when the functions of a Central Statistical Board, created in 1933, were transferred to the then-named Bureau of the Budget (see Anderson, 1988; Duncan and Shelton, 1978; Norwood, 1995). Recent legislation and guidance addresses such system-wide issues as confidentiality protection and privacy of respondents, data quality (including peer review prior to dissemination), efficiency of operations, and evaluation of agencies' performance.

THE PAPERWORK REDUCTION ACT AND ASSOCIATED GUIDANCE

The 1980 Paperwork Reduction Act (PRA) (Title 44, Section 35 of the U.S. Code, amended in 1986 and reauthorized in 1995 by P.L. 104-13) is the foundation for the modern statistical coordination and management mission of the Office of Management and Budget. It establishes OMB's review power not only over federal statistical agencies, but also over the myriad other agencies throughout the federal government that collect information from individuals and organizations. This review power covers both data collection budgets and methods and practices for data collection and dissemination.

Background

The PRA's origins trace back to Executive Order 6226, signed by Franklin D. Roosevelt in July 1933, which established a Central Statistical Board to "appraise and advise upon all schedules of all Government agencies engaged in the primary collection of statistics required in carrying out the purposes of the National Industrial Recovery Act, to review plans for tabulation and classification of such statistics, and to promote the coordination and improvement of the statistical services involved." Members of the board were appointed by relevant cabinet secretaries. The board was established in law for a 5-year period in 1935. Its functions were transferred to the Bureau of the Budget (itself established in 1921) in 1939, when the Budget Bureau was transferred to the Executive Office of the President.

The 1942 Federal Reports Act represented another milestone: It codified the authority for the Budget Bureau to coordinate and oversee the work of federal statistical agencies. Most famously, it provided that no federal agency could collect data from 10 or more respondents without approval of the budget director. (Data collections by contractors on behalf of federal agencies are covered by this provision, although data collections by government grantees are generally not covered.) The 1950 Budget and Accounting Procedures Act (Title 31, Section 1104(d) of the U.S. Code) further strengthened the statistical coordinating and improvement role of OMB, including giving OMB authorization to promulgate regulations and orders governing statistical programs throughout the federal government.

The statistical policy function continued in the budget office in the Executive Office of the President when the Budget Bureau became the Office of Management and Budget in 1970. However, in 1977, the statistical policy staff was split into two groups—one group remained with OMB to handle the paperwork clearance and review function for statistical agencies; the other group was moved to the Department of Commerce to address statistical policy and standards issues.

Paperwork Reduction

The overarching goal of the 1980 Paperwork Reduction Act was to reduce the burden of filling out federal forms by businesses and individuals. It moved the statistical policy office back to OMB (from the Department of Commerce) under a new Office of Information and Regulatory Affairs (OIRA), which was charged to reduce the combined burden imposed by regulatory agencies and statistical agencies. The PRA required OMB, through the chief statistician, to engage in long-range planning to improve federal statistical programs; review statistical budgets; coordinate government statistical functions; establish standards, classifications, and other guidelines for statistical data collection and dissemination; and evaluate statistical program performance. In the 1995 reauthorization and extensive revision of PRA, the most important provision for statistical policy was to authorize the establishment of the Interagency Council on Statistical Policy (ICSP), chaired by the chief statistician, who heads a small staff in the Statistical and Science Policy Office (see Appendix A).

Survey Clearance Process

The OMB Statistical and Science Policy Office released in January 2006 *Guidance on Agency Survey and Statistical Information Collections—Questions and Answers When Designing Surveys for Information Collections* (see http://www.whitehouse.gov/omb/inforeg/pmc_survey_guidance_2006.pdf).

It is a set of 81 questions and answers that attempts to demystify the OMB clearance process (required by the PRA) for surveys and other statistical information collections. Its purpose is to explain OMB's review process, assist in strengthening supporting statements for information collection requests, and provide advice for improving information collection designs.

The *Guidance* covers such topics as the purpose of the guidance; submission of information collection requests (often called clearance packages) to OMB; scope of the information collection (e.g., calculation of burden hours on respondents); choice of methods; sampling; modes of data collection; questionnaire design and development; statistical standards; informing respondents about their participation and the confidentiality of their data; response rates and incentives; analysis and reporting; and studies using stated preference methods (which ask respondents about the use or non-use value of a good in order to obtain willingness-to-pay estimates relevant to benefit or cost estimation). The guidance includes a glossary of terms and examples of the OMB clearance request forms.

The guidance outlines the timing and process requirements for all statistical information collection requests in order to obtain OMB approval (which is indicated by an OMB control number on approved survey questionnaires). First, the agency must publish a 60-day notice in the *Federal Register* inviting public comment on the proposed collection. At this stage, the agency must have at least a draft survey instrument for the public for review. Following this initial comment period, the agency may submit its clearance package to OMB, at which time it must place a second notice in the *Federal Register*, allowing a 30-day public comment period and notifying the public that OMB approval is being sought and that comments may be submitted to OMB. This notice runs concurrent with the first 30 days of OMB review, and OMB has a total of 60 days after receipt of the clearance package to make its decision to approve or disapprove the package or to instruct the agency to make a substantive change to its proposed collection. Generally, agencies need to allow 6 months to complete the entire process, including agency and departmental review.

2000 INFORMATION QUALITY ACT AND ASSOCIATED GUIDELINES

The Information Quality Act (IQA) of 2000 requires federal agencies to develop procedures that will ensure the quality of information they disseminate and to develop an administrative mechanism whereby affected parties can request that agencies correct low-quality information. The act directs OMB to issue guidelines "ensuring and maximizing the quality, objectivity, utility, and integrity of information . . . disseminated by Federal agencies."

OMB first published proposed government-wide IQA guidelines in the *Federal Register* on June 28, 2001 (66 *Federal Register* 34489). It issued final guidelines (after incorporating changes pursuant to public comments) on February 22, 2002 (67 *Federal Register* 8452-8460) (see also http://www.whitehouse.gov/omb/inforeg/iqg_2002.pdf). A few months later, 13 statistical agencies issued a *Federal Register* notice (67 *Federal Register* 38467-38470, June 4) outlining a common approach to the development and provision of guidelines for ensuring and maximizing the quality, objectivity, utility, and integrity of disseminated information. The notice directed people to the websites of each agency for more information and to learn how to comment on draft guidelines. Each agency then finalized its own guidelines (see, e.g., http://www.census.gov/qdocs/www/quality_guidelines.htm). The framework developed by the agencies was followed in the 2006 revision of OMB's standards and guidelines for statistical surveys (see OMB Statistical Policy Directives below).

1997 ORDER PROVIDING FOR THE CONFIDENTIALITY OF STATISTICAL INFORMATION

OMB issued the *Order Providing for the Confidentiality of Statistical Information* in 1997 (*Federal Register*, Vol. 62, No. 124, Friday, June 27, 1997; see http://www.whitehouse.gov/omb/inforeg/conf-order.pdf). The order was designed to bolster the confidentiality protections afforded by statistical agencies or units (as listed in the order), some of which lacked legal authority to back up their confidentiality protection. CIPSEA (see below) placed confidentiality protection for statistical information on a strong legal footing across the entire federal government.

2002 CONFIDENTIAL INFORMATION PROTECTION AND STATISTICAL EFFICIENCY ACT AND ASSOCIATED GUIDANCE

The Confidential Information Protection and Statistical Efficiency Act (CIPSEA) (Title V of the E-Government Act of 2002, P.L. 107-347) was landmark legislation to strengthen the statistical system with regard to confidentiality protection and data sharing. Enactment of CIPSEA culminated more than 30 years of efforts to standardize and bolster legal protections for data collected for statistical purposes by federal agencies while permitting limited sharing of individually identifiable business information among three statistical agencies for efficiency and quality improvement. CIPSEA has two subtitles, A and B, covering confidentiality and sharing data.

Subtitle A—Protecting Confidentiality

Subtitle A, Confidential Information Protection, strengthens and extends statutory confidentiality protection for all statistical data collections of the U.S. government. Prior to CIPSEA, such protection was governed by a patchwork of laws applicable to specific agencies, judicial opinions, and agency practice. For all data furnished by individuals or organizations to an agency under a pledge of confidentiality for exclusively statistical purposes, Subtitle A provides that the data will be used only for statistical purposes and will not be disclosed in identifiable form to anyone not authorized by the title. It makes knowing and willful disclosure of confidential statistical data a class E felony with fines up to $250,000 and imprisonment for up to 5 years.

Subtitle A pertains not only to surveys, but also to collections by a federal agency for statistical purposes from administrative records (e.g., state government agency records). Data covered under Subtitle A are not subject to release under a Freedom of Information Act request.

Subtitle B—Sharing Data

Subtitle B of CIPSEA, Statistical Efficiency, permits the Bureau of Economic Analysis (BEA), the Bureau of Labor Statistics (BLS), and the Census Bureau to share individually identifiable business data for statistical purposes. The intent of the subtitle is to reduce respondent burden on businesses; improve the comparability and accuracy of federal economic statistics by permitting these three agencies to reconcile differences among sampling frames, business classifications, and business reporting; and increase understanding of the U.S. economy and improve the accuracy of key national indicators, such as the National Income and Product Accounts.

Several data-sharing projects have been initiated under Subtitle B. However, it does not permit sharing among BEA, BLS, and the Census Bureau of any individually identifiable tax return data that originates from the Internal Revenue Service (IRS). This limitation currently blocks some important kinds of business data sharing for improving the efficiency and quality of business data collection by statistical agencies.

For tax return information, data sharing is limited to a small number of items for specified uses by a small number of specific agencies (under Title 26, Section 6103 of the U.S. Code and associated Treasury Department regulations, as modified in the 1976 Tax Reform Act). The law provides access to specific tax return items by the Census Bureau for use in its population estimates program and economic census and survey programs, by the National Agricultural Statistics Service for conducting the Census of Agriculture, by the Congressional Budget Office for evaluation of government revenue and expenditure initiatives, and by BEA for producing the National Income and Product Accounts. (Prior to the act, the President could issue an Executive Order authorizing access to tax records.) The governing statute would have to be modified to extend sharing of tax return items to agencies not specified in the 1976 legislation.

A proposal for legislation was developed through interagency discussions in 2008 that would authorize the Bureau of Labor Statistics to receive limited business data from the Census Bureau (comingled with business tax information) for the purpose of synchronizing the two agencies' business lists. It would also authorize the Bureau of Economic Analysis to receive business tax information for noncorporate businesses with receipts greater than $1 million, allowing BEA to improve the measurement of income and international transactions in the national accounts.

OMB CIPSEA Guidance

OMB is charged to oversee and coordinate the implementation of CIPSEA; after a long interagency coordination process, OMB released final guidance in June 2007 (see http://www.whitehouse.gov/omb/fedreg/2007/061507_cipsea_guidance.pdf). The guidance, which pertains to both Subtitles A and B, covers such topics as the steps that agencies must take to protect confidential information; wording of confidentiality pledges in materials that are provided to respondents; steps that agencies must take to distinguish any data or information they collect for nonstatistical purposes and to provide proper notice to the public of such data; and ways in which agents (e.g., contractors, researchers) may be designated to use individually identifiable information for analysis and other statistical purposes and be held legally responsible for protecting the confidentiality of that information.

A key provision of the CIPSEA guidance defines statistical agencies or units, which are the only federal agencies that may assign agent status for confidentiality protection purposes to contractors, researchers, or others. The guidance defines a statistical agency or unit as "an agency or organizational unit of the executive branch whose activities are predominantly the collection, compilation, processing, or analysis of information for statistical purposes." Twelve statistical agencies that were enumerated in OMB's 1997 Confidentiality Order are recognized in OMB's CIPSEA implementation guidance as statistical agencies for the purposes of CIPSEA: the members of the Interagency Council on Statistical Policy (ICSP) (see Appendix A), excluding the Office of Environmental Information in the Environmental Protection Agency and the Office of Research, Evaluation, and Statistics in the Social Security Administration. During 2007, OMB recognized two more statistical units that applied for designation under the procedures outlined in the guidance: the Office of Applied Studies in the Substance Abuse and Mental Health Services Administration in the Department of Health and Human Services, and the Microeconomic Surveys Section of the Board of Governors of the Federal Reserve.

2002 E-GOVERNMENT ACT, SECTION 208

Section 208 of the E-Government Act of 2002 requires federal agencies to conduct a privacy impact assessment whenever the agency develops or obtains information technology that handles individually identifiable information or whenever the agency initiates a new collection of individually identifiable information.[1] The assessment is to be made publicly available and cover such topics as what information is being collected and why, with whom the information will be shared, what provisions will be made for informed consent regarding data sharing, and how the information will be

[1]Section 208 also mandates that OMB lead interagency efforts to improve federal information technology (IT) and use of the Internet for government services.

secured. Typically, privacy impact assessments cover not only privacy issues, but also confidentiality, integrity, and availability issues (see, e.g., U.S. Census Bureau, 2003). OMB is required to issue guidance for development of the assessments, which was done in a memorandum from the OMB director to the heads of executive agencies and departments on September 26, 2003 (see http://www.whitehouse.gove/omb/memoranda/m03-22html). Section 208, Title III (see below), and Title V (see above) are the latest in a series of laws dating back to 1974 that govern access to individual records maintained by the federal government. The Privacy Act of 1974 states in part:

> No agency shall disclose any record which is contained in a system of records by any means of communication to any person, or to another agency, except pursuant to a written request by, or with the prior written consent of, the individual to whom the record pertains. . . .

There are specific exceptions allowing the use of personal records without prior consent for statistical purposes by the Census Bureau, for statistical research or reporting when the records are to be transferred in a form that is not individually identifiable, for routine uses within a U.S. government agency, for archival purposes "as a record which has sufficient historical or other value to warrant its continued preservation by the United States Government," for law enforcement purposes, for congressional investigations, and for other administrative purposes. The Privacy Act mandates that every federal agency have in place an administrative and physical security system to prevent the unauthorized release of personal records.

2002 FEDERAL INFORMATION SECURITY MANAGEMENT ACT

The Federal Information Security Management Act (FISMA) was enacted in 2002 as Title III of the E-Government Act of 2002 (P.L. 107-347). The act was meant to bolster computer and network security within the federal government and affiliated parties (such as government contractors) by mandating yearly audits.

FISMA imposes a mandatory set of processes that must be followed for all information systems used or operated by a federal agency or by a contractor or other organization on behalf of a federal agency. These processes must follow a combination of Federal Information Processing Standards (FIPS) documents, the special publications SP-800 series issued by the National Institute of Standards and Technology, and other legislation pertinent to federal information systems, such as the Privacy Act of 1974 and the Health Insurance Portability and Accountability Act.

The first step is to determine what constitutes the "information system" in question. There is not a direct mapping of computers to an information system; rather an information system can be a collection of individual computers put to a common purpose and managed by the same system owner. The next step is to determine the types of information resident in the system and categorize each according to the magnitude of harm resulting were the system to suffer a compromise of confidentiality, integrity, or availability. Succeeding steps are to develop complete system documentation, conduct a risk assessment, put appropriate controls in place to minimize risk, and arrange for an assessment and certification of the adequacy of the controls.

FISMA affects federal statistical agencies directly in that each of them must follow the FISMA procedures for its own information system. In addition, some departments are taking the position that all information systems within a department constitute a single information system for purposes of FISMA. As a consequence, these departments are taking steps to require that statistical agencies' information systems and personnel be incorporated into a centralized department-wide system.

2004 OMB PEER REVIEW GUIDANCE

In 2003-2004, under the authority of the 2000 Information Quality Act, OMB developed guidance for federal agencies with regard to seeking peer review of policy-related information an agency disseminates. The *Final Information Quality Bulletin for Peer Review* was issued on December 16, 2004; it requires federal agencies to conduct a peer review of "influential scientific information" before the information is released to the public (see http://www.whitehouse.gov/omb/memoranda/fy2005/m05-03.pdf). "Influential scientific information" is defined as "scientific information the agency reasonably can determine will have or does have a clear and substantial impact on important public policies or private sector decisions" (U.S. Office of Management and Budget, 2004a:10). The bulletin grants agencies discretion to select the type of peer review process for a given information product. The bulletin excludes from the guidelines "routine statistical information released by federal statistical agencies (e.g., periodic demographic and economic statistics) and the analysis of these data to compute standard indicators and trends (e.g., unemployment and poverty rates)" (U.S. Office of Management and Budget, 2004a:40).

2002-2008 PERFORMANCE ASSESSMENT RATING TOOL

The Office of Management and Budget began a major initiative in 2002 to develop a tool for assessing the performance of federal agencies and programs that would identify effective and ineffective programs and provide information that could be used in making budgetary decisions. The tool is the Performance Assessment Rating Tool (PART), which has sets of questions or measures on program design and purpose, program goals, agency management of programs, and results that agencies can report with accuracy and consistency. (There are several versions of the PART for use by different kinds of programs, such as research and development programs, competitive grant programs, or direct federal programs.) Answers to the questions in each section produce a numeric score for that section from 0 to 100; the section scores are then combined to achieve an overall qualitative rating: Effective, Moderately Effective, Adequate, Ineffective, or Results Not Demonstrated (for more information, see http://www.whitehouse/gov/omb/part/2004_fax.html).

The members of the ICSP collaborated to develop an initial set of common performance standards for use in completing PART and in developing strategic plans required by the Government Performance and Results Act of 1993 (GPRA). The agencies agreed on two general areas of focus—product quality and program performance—and on three dimensions of each focus area. For product quality, the dimensions are relevance, accuracy, and timeliness; for program performance, the dimensions are cost, dissemination, and mission achievement. Example indicators were developed for each dimension, such as measures of customer satisfaction as an indicator of mission achievement (see U.S. Office of Management and Budget, 2008a:37-41).

OMB STATISTICAL POLICY DIRECTIVES

The Statistical and Science Policy Office issues and periodically updates a number of directives that pertain to federal agency data collection and dissemination. The oldest two directives on standards for statistical surveys and publication of statistics, first issued in the 1950s and updated in the 1970s, were recently combined as part of a major overhaul. A 1969 directive on the official poverty measure was updated in a minor way in 1978; the standards on Metropolitan Statistical Areas and industry and occupation classifications are updated at least every 10 years; the newest directive on release of statistical products, issued in 2008, complements a directive issued in the 1970s and updated in 1985 on release of economic indicators. Among the statistical policy directives are the following, which are briefly summarized below:[2]

- Standards and Guidelines for Statistical Surveys (replaces and combines Statistical Policy Directives Nos. 1 and 2)
- Statistical Policy Directive No. 3—Compilation, Release, and Evaluation of Principal Federal Economic Indicators (and Schedule of Release Dates for Principal Federal Economic Indicators)
- Statistical Policy Directive No. 4—Release and Dissemination of Statistical Products Produced by Federal Statistical Agencies
- Metropolitan Statistical Areas
- North American Industry Classification System (NAICS)/Standard Industrial Classification (SIC)
- Standard Occupational Classification (SOC)
- Statistical Policy Directive No. 14—Definition of Poverty for Statistical Purposes
- Standards for Maintaining, Collecting, and Presenting Federal Data on Race and Ethnicity

Standards and Guidelines for Statistical Surveys

OMB issued *Standards and Guidelines for Statistical Surveys* in September 2006 (see http://www.whitehouse.gov/omb/inforeg/statpolicy/standards_stat_surveys.pdf) as an update and revision of *Statistical Policy Directive No. 1, Standards for Statistical Surveys,* and *Statistical Policy Directive No. 2, Publication of Statistics* (see http://www.whitehouse.gov//omb/fedreg/2006/09226_stat_surveys.html; see also http://www.whitehouse.gov/omb/inforeg/backgrd_stat_surveys.html). The new document includes 20 standards and one or more associated guidelines for every aspect of survey methodology from planning through data release:

(1) Survey planning.
(2) Survey design.

---

[2]All of the directives are available at or can be linked to from http://www.whitehouse.gov/omb/inforeg/statpolicy.html, with the exception of Statistical Policy Directive No. 14, which is available at http://www.census.gov/hhes/www/povmeas/ombdir14.html.

(3) Survey response rates.
(4) Pretesting survey systems.
(5) Developing sampling frames.
(6) Required notification to potential survey respondents.
(7) Data collection methodology.
(8) Data editing.
(9) Nonresponse analysis and response rate calculation.
(10) Coding.
(11) Data protection.
(12) Evaluation.
(13) Developing estimates and projections.
(14) Analysis and report planning.
(15) Inference and comparisons.
(16) Review of information products.
(17) Releasing information.
(18) Data protection and disclosure avoidance for dissemination.
(19) Survey documentation.
(20) Documentation and release of public-use microdata.

Principal Economic Indicators

OMB issued *Statistical Policy Directive No. 3—Compilation, Release, and Evaluation of Principal Federal Economic Indicators* in the 1970s and strengthened it in 1985 (*Federal Register*, Vol. 50, No. 186, Wednesday, September 25, 1985; see http://www.whitehouse.gov/omb/inforeg/statpolicy/dir_3_fr_09251985.pdf). Its purpose is clearly stated:

> [This directive] designates statistical series that provide timely measures of economic activity as Principal Economic Indicators and requires prompt release of these indicators by statistical agencies in a politically-neutral manner. The intent of the directive is to preserve the time value of such information, strike a balance between timeliness and accuracy, prevent early access to information that may affect financial and commodity markets, and preserve the distinction between the policy-neutral release of data by statistical agencies and their interpretation by policy officials.

Each September OMB issues the *Schedule of Release Dates for Principal Federal Economic Indicators* for the subsequent calendar year (see http://www.whitehouse.gov/omb/inforeg/statpolicy.html#sr). At present, the following agencies issue one or more principal economic indicators (38 indicators in all):

- Bureau of Economic Analysis (5 indicators, including gross domestic product, personal income and outlays, corporate profits).
- Bureau of Labor Statistics (7 indicators, including the employment situation, Consumer Price Index).
- Census Bureau (13 indicators, including new residential construction, monthly wholesale trade).
- Energy Information Administration (weekly natural gas storage report).
- Federal Reserve Board (4 indicators, including money stock measures, consumer installment credit).
- Foreign Agricultural Service (world agricultural production).
- National Agricultural Statistics Service (6 indicators, including agricultural prices and grain production).
- World Agricultural Outlook Board (world agricultural supply and demand estimates).

Release and Dissemination of Statistical Products

OMB issued *Statistical Policy Directive No. 4—Release and Dissemination of Statistical Products Produced by Federal Statistical Agencies* in 2008 (*Federal Register*, Vol. 73, No. 46, Friday, March 7, 2008) as a companion to Directive 3 (see http://www.whitehouse.gov/omb/fedreg/2008/030708_directive-4.pdf). Directive 4 essentially covers all statistical releases other than those specified in Directive 3. It includes not only statistical information released in printed reports or on the Internet, but also statistical press releases, which describe or announce a statistical data product. Statistical press releases are the sole responsibility of the relevant statistical agency. Each fall statistical agencies must issue a schedule of when they expect each regular or recurring product to be released and give timely notification of any change to the published schedule.

North American Industry Classification System

The North American Industry Classification System (NAICS) updates and substantially revises the old Standard Industrial Classification (SIC) (see http://www. census.gov/epcd/www/naics.html). The NAICS was developed by the United States, Canada, and Mexico to provide a common, contemporary classification system for economic production activity following on the enactment of the North American Free Trade Agreement (NAFTA).

Interagency and country working groups (under the aegis of OMB in the United States) are updating the NAICS every 5 years so that it keeps up reasonably well with changes in industrial activity in the three countries.

Standard Occupational Classification

The Standard Occupational Classification (SOC) is used by federal statistical agencies to classify workers into occupational categories for collecting, calculating, and disseminating data (see http://www.bls.gov/soc/). Historically, the SOC has been revised prior to each decennial census through an interagency process under the aegis of OMB and has been used in surveys conducted during the following decade. With the advent of the American Community Survey (which provides occupational data in place of the decennial census "long-form" sample), future timing of SOC revisions is under study. Work to revise the 2000 Standard Occupational Classification for 2010 should be completed shortly.

Metropolitan Area Classification

For more than 50 years, the OMB Metropolitan Area Classification Program has provided standard statistical area definitions for use throughout the federal government. The usefulness of standardizing these classifications became clear in the 1940s, and the Bureau of the Budget (the predecessor to OMB) led an effort to develop what were then called "standard metropolitan areas" in time for their use in 1950 census publications. Since then, OMB has updated as appropriate the definitional criteria for metropolitan areas before each census; based on those criteria, after each census, OMB has issued a list of recognized areas.

The definitional criteria issued before the 2000 census marked a major revision to the coverage of the program. *Standards for Defining Metropolitan and Micropolitan Statistical Areas* (*Federal Register*, Vol. 65, No. 249, Wednesday, December 27, 2000) defined not only metropolitan statistical areas, but also, for the first time, micropolitan areas (see http://www.whitehouse.gov/omb/fedreg/metroareas122700.pdf). Metropolitan areas are those with a central urbanized core of 50,000 or more people in one or more counties; micropolitan areas are those with a central urbanized core of 10,000 or more people in one or more counties. The list of metropolitan and micropolitan areas is annually updated by OMB on the basis of the Census Bureau's population estimates (see, e.g., *Update of Statistical Area*

Definitions and Guidance on Their Uses, issued November 20, 2008; http://www.whitehouse.gov/omb/bulletins/fy2009/09-01.pdf).

Definition of Poverty

OMB first issued *Statistical Policy Directive No. 14—Definition of Poverty for Statistical Purposes* in 1969. The directive adopted the poverty thresholds first defined by Mollie Orshansky of the Social Security Administration for 1963 for different categories of families defined by size, number of children, gender of the family head, and farm-nonfarm residences. It specified that these thresholds would be updated each year for the change in the Consumer Price Index and compared with families' total money income as measured in the Current Population Survey. The directive was reissued in 1978; additional minor modifications were made to the poverty thresholds beginning in 1982 (see http://www.census.gov/hhes/www/poverty/prevcps/p60-133.pdf#page=9).

Data on Race and Ethnicity

OMB first issued *Statistical Policy Directive No. 15—Race and Ethnic Standards for Federal Statistics and Administrative Reporting* in 1977. It specified a minimum set of racial and ethnic categories for reporting of race and ethnicity on federal surveys and in administrative records systems. It recommended two separate questions—one on ethnicity (Hispanic or non-Hispanic) and one on race (white, black, Asian or Pacific Islander, American Indian or Alaska Native), or, alternatively, a combined question that included Hispanic as a category. The U.S. census has historically included additional categories within the two-question format. Following an intensive research, testing, and consultation process, OMB issued revised *Standards for Maintaining, Collecting, and Presenting Federal Data on Race and Ethnicity* on October 30, 1997 (see http://www.whitehouse.gov/omb/fedreg/1997standards.html). The updated directive retains a two-question format, includes separate categories for Asians and for Native Hawaiian and other Pacific Islanders, and allows respondents to select more than one racial category.

Appendix C

Fundamental Principles of Official Statistics of the Statistical Commission of the United Nations

[The text below is excerpted from the report of the Statistical Commission on its Special Session, held in New York City, April 11-15, 1994. Official Records of the Economic and Social Council, 1994, Supplement No. 9.]

Action taken by the Commission

59. The Commission adopted the fundamental principles of official statistics as set out in ECE decision C (47), but incorporating a revised preamble. The preamble and principles, as adopted, are set out below:

FUNDAMENTAL PRINCIPLES OF OFFICIAL STATISTICS

The Statistical Commission,

Bearing in mind that official statistical information is an essential basis for development in the economic, demographic, social and environmental fields and for mutual knowledge and trade among the States and peoples of the world,

Bearing in mind that the essential trust of the public in official statistical information depends to a large extent on respect for the fundamental values and principles which are the basis of any society which seeks to understand itself and to respect the rights of its members,

Bearing in mind that the quality of official statistics, and thus the quality of the information available to the Government, the economy and the public depends largely on the cooperation of citizens, enterprises, and other respondents in providing appropriate and reliable data needed for necessary statistical compilations and on the cooperation between users and producers of statistics in order to meet users' needs,

Recalling the efforts of governmental and non-governmental organizations active in statistics to establish standards and concepts to allow comparisons among countries,

Recalling also the International Statistical Institute Declaration of Professional Ethics,

Having expressed the opinion that resolution C (47), adopted by the Economic Commission for Europe on 15 April 1992, is of universal significance,

Noting that, at its eighth session, held at Bangkok in November 1993, the Working Group of Statistical Experts, assigned by the Committee on Statistics of the Economic and Social Commission for Asia and the Pacific to examine the Fundamental Principles, had agreed in principle to the ECE version and had emphasized that those principles were applicable to all nations,

Noting also that, at its eighth session, held at Addis Ababa in March 1994, the Joint Conference of African Planners, Statisticians and Demographers, considered that the Fundamental Principles of Official Statistics are of universal significance,

Adopts the present principles of official statistics:

1. Official statistics provide an indispensable element in the information system of a democratic society, serving the Government, the economy and the public with data about the economic, demographic, social and environmental situation. To this end, official statistics that meet the test of practical utility are to be compiled and made available on an impartial basis by official statistical agencies to honour citizens' entitlement to public information.

2. To retain trust in official statistics, the statistical agencies need to decide according to strictly professional considerations, including scientific principles and professional ethics, on the methods and procedures for the collection, processing, storage and presentation of statistical data.

3. To facilitate a correct interpretation of the data, the statistical agencies are to present information according to scientific standards on the sources, methods and procedures of the statistics.

4. The statistical agencies are entitled to comment on erroneous interpretation and misuse of statistics.

5. Data for statistical purposes may be drawn from all types of sources, be they statistical surveys or administrative records. Statistical agencies are to choose the source with regard to quality, timeliness, costs and the burden on respondents.

6. Individual data collected by statistical agencies for statistical compilation, whether they refer to natural or legal persons, are to be strictly confidential and used exclusively for statistical purposes.

7. The laws, regulations and measures under which the statistical systems operate are to be made public.

8. Coordination among statistical agencies within countries is essential to achieve consistency and efficiency in the statistical system.

9. The use by statistical agencies in each country of international concepts, classifications and methods promotes the consistency and efficiency of statistical systems at all official levels.

10. Bilateral and multilateral cooperation in statistics contributes to the improvement of systems of official statistics in all countries.

Appendix D

Selected Federal Statistical Websites, January 2009

The information in this appendix is adapted from a listing in U.S. Office of Management and Budget (2008c); it includes both federal statistical agencies and other federal agencies that produce statistical information.

Congressional Budget Office (CBO): www.cbo.gov

Consumer Product Safety Commission (CPSC): www.cpsc.gov

Department of Agriculture (USDA): www.usda.gov
- Agricultural Research Service (ARS): www.ars.usda.gov
- Economic Research Service (ERS): www.ers.usda.gov
- Food and Nutrition Service (FNS): www.fns.usda.gov
- Foreign Agricultural Service (FAS): www.fas.usda.gov
- Forest Service (FS): www.fs.fed.us
- National Agricultural Statistics Service (NASS): www.usda.gov/nass
- Natural Resources Conservation Service (NRCS): www.nrcs.usda.gov

Department of Commerce: www.doc.gov
- Bureau of Economic Analysis (BEA): www.bea.gov
- Census Bureau: www.census.gov
- Economics and Statistics Administration (ESA): www.esa.doc.gov
- International Trade Administration (ITA): www.ita.doc.gov

National Environmental Satellite, Data, and Information Service: www.nesdis.noaa.gov
National Marine Fisheries Service (NMFS): www.nmfs.noaa.gov
National Oceanic and Atmospheric Administration (NOAA): www.noaa.gov

Department of Defense: www.dod.gov
Army Corps of Engineers (CORPS): www.usace.army.mil
Defense Manpower Data Center (DMDC): www.dmdc.osd.mil

Department of Education: www.ed.gov
National Center for Education Statistics (NCES): www.nces.ed.gov

Department of Energy: www.doe.gov
Energy Information Administration (EIA): www.eia.doe.gov
Office of Health, Safety, and Security (HSS): www.hss.doe.gov

Department of Health and Human Services: www.hhs.gov
Administration for Children and Families (ACF): www.acf.hhs.gov
Agency for Healthcare Research and Quality (AHRQ): www.ahrq.gov
Agency for Toxic Substances and Disease Registry (ATSDR): www.atsdr.cde.gov
Centers for Disease Control and Prevention (CDC): www.cdc.gov
Centers for Medicare and Medicaid Services (CMS): www.cms.hhs.gov
Health Resources and Services Administration (HRSA): www.hrsa.gov
HHS Data Council: www.hhs-stat.net
Indian Health Service (IHS): www.ihs.gov
National Center for Health Statistics (NCHS): www.cdc.gov/nchs
National Institute of Child Health and Human Development (NICHD): www.nichd.nih.gov
National Institute on Aging (NIA): www.nia.nih.gov
National Institutes of Health (NIH): www.nih.gov
Office of Population Affairs (OPA): www.opa.osophs.dhhs.gov
Office of the Assistant Secretary for Planning and Evaluation (ASPE): www.aspe.hhs.gov
Substance Abuse and Mental Health Services Administration, Office of Applied Studies (SAMHSA/OAS): www.oas.samhsa.gov

Department of Homeland Security: www.dhs.gov
- Bureau of Customs and Border Protection: www.cbp.gov
- Federal Emergency Management Agency (FEMA): www.fema.gov
- Office of Immigration Statistics (OIS): www.dhs.gov/ximgtn/statistics

Department of Housing and Urban Development: www.hud.gov
- Office of Federal Housing Enterprise Oversight (OFHEO): www.ofheo.gov
- Office of Public and Indian Housing (PIH): www.hud.gov/offices/pih/index.cfm
- Office of the Assistant Secretary for Policy Development and Research (PD&R): www.huduser.org

Department of the Interior: www.doi.gov
- Bureau of Reclamation (BoR): www.usbr.gov
- Minerals Management Service (MMS): www.mms.gov
- National Park Service (NPS): www.nps.gov
- United States Fish and Wildlife Service (FWS): www.fws.gov
- United States Geological Survey (USGS): www.usgs.gov

Department of Justice: www.usdoj.gov
- Bureau of Justice Statistics (BJS): www.ojp.usdoj.gov/bjs
- Bureau of Prisons (BoP): www.bop.gov
- Drug Enforcement Administration (DEA): www.usdoj.gov/dea
- Federal Bureau of Investigation (FBI): www.fbi.gov

Department of Labor: www.dol.gov
- Bureau of Labor Statistics (BLS): www.bls.gov
- Employment and Training Administration (ETA): www.doleta.gov
- Employment Standards Administration (ESA): www.dol.gov/esa
- Mine Safety and Health Administration (MSHA): www.msha.gov
- Occupational Safety and Health Administration (OSHA): www.osha.gov

Department of Transportation: www.dot.gov
- Bureau of Transportation Statistics (BTS): www.bts.gov
- Federal Aviation Administration (FAA): www.faa.gov
- Federal Highway Administration (FHWA): www.fhwa.dot.gov

Federal Motor Carrier Safety Administration (FMCSA): www.fmcsa.dot.gov
Federal Railroad Administration (FRA): www.fra.dot.gov
Federal Transit Administration (FTA): www.fta.dot.gov
Maritime Administration (MARAD): www.marad.dot.gov
National Highway Traffic Safety Administration (NHTSA): www.nhtsa.dot.gov/people/ncsa
Pipeline and Hazardous Materials Safety Administration (PHMSA): www.phmsa.dot.gov

Department of the Treasury: www.ustreas.gov
Internal Revenue Service (IRS): www.irs.treas.gov
Statistics of Income (SOI): www.irs.gov/taxstats/index.html

Department of Veterans Affairs (VA): www.va.gov
National Center for Veterans Analysis and Statistics (NCVAS): www.va.gov/vetdata

Environmental Protection Agency (EPA): www.epa.gov
Office of Environmental Information: www.epa.gov/oei

Equal Employment Opportunity Commission (EEOC): www.eeoc.gov

Executive Office of the President
Federal Statistics Briefing Rooms
Economic Statistics (ESBR): www.whitehouse.gov/fsbr/esbr.html
Social Statistics (SSBR): www.whitehouse.gov/fsbr/ssbr.html
Office of Management and Budget (OMB): www.whitehouse.gov/omb (Go to "Statistical Programs and Standards")
Federal Committee on Statistical Methodology: www.fcsm.gov
Federal Interagency Council on Statistical Policy, Federal Statistics: www.fedstats.gov
Federal Interagency Forum on Aging-Related Statistics: www.agingstats.gov
Federal Interagency Forum on Child and Family Statistics: www.childstats.gov

Federal Reserve Board: www.federalreserve.gov

National Aeronautics and Space Administration (NASA): www.nasa.gov

National Science Foundation (NSF): www.nsf.gov
Directorate for Social, Behavioral, and Economic Sciences: www.nsf.gov/sbe
Methodology, Measurement, and Statistics Program (MMS): www.nsf.gov/sbe/ses/mms/start.htm
Science Resources Statistics Division (SRS): www.nsf.gov/sbe/srs/

Small Business Administration (SBA): www.sba.gov

Social Security Administration (SSA): www.socialsecurity.gov
Office of Research, Evaluation, and Statistics (ORES): www.socialsecurity/policy

U.S. Agency for International Development (USAID): www.usaid.gov

U.S. Government Accountability Office (GAO): www.gao.gov

INDEX OF FEDERAL STATISTICAL SITES LISTED

Census Bureau—see Department of Commerce
Centers for Disease Control and Prevention—see Department of Health and Human Services
Centers for Medicare and Medicaid Services—see Department of Health and Human Services
Congressional Budget Office—see Congressional Budget Office
Consumer Product Safety Commission—see Consumer Product Safety Commission
Defense Manpower Data Center—see Department of Defense
Directorate for Social, Behavioral, and Economic Sciences—see National Science Foundation
Drug Enforcement Administration—see Department of Justice
Economic Research Service—see Department of Agriculture
Economic Statistics Briefing Room—see Executive Office of the President, Federal Statistics Briefing Rooms
Economics and Statistics Administration—see Department of Commerce
Employment and Training Administration—see Department of Labor
Energy Information Administration—see Department of Energy
Environmental Protection Agency—see Environmental Protection Agency
Federal Aviation Administration—see Department of Transportation
Federal Bureau of Investigation—see Department of Justice
Federal Committee on Statistical Methodology—see Executive Office of the President, Office of Management and Budget
Federal Emergency Management Agency—see Department of Homeland Security
Federal Highway Administration—see Department of Transportation
Federal Interagency Council on Statistical Policy, Federal Statistics—see Executive Office of the President, Office of Management and Budget
Federal Interagency Forum on Aging-Related Statistics—see Executive Office of the President, Office of Management and Budget
Federal Interagency Forum on Child and Family Statistics—see Executive Office of the President, Office of Management and Budget
Federal Motor Carrier Safety Administration—see Department of Transportation
Federal Railroad Administration—see Department of Transportation
Federal Statistics Briefing Rooms—see Executive Office of the President
Federal Transit Administration—see Department of Transportation
Food and Nutrition Service—see Department of Agriculture

Foreign Agricultural Service—see Department of Agriculture
Forest Service—see Department of Agriculture
Health Resources and Services Administration—see Department of Health and Human Services
Indian Health Service—see Department of Health and Human Services
Internal Revenue Service—see Department of the Treasury
International Trade Administration—see Department of Commerce
Maritime Administration—see Department of Transportation
Methodology, Measurement, and Statistics Program—see National Science Foundation, Directorate for Social, Behavioral, and Economic Sciences
Mine Safety and Health Administration—see Department of Labor
Minerals Management Service—see Department of the Interior
National Aeronautics and Space Administration—see National Aeronautics and Space Administration
National Agricultural Statistics Service—see Department of Agriculture
National Center for Education Statistics—see Department of Education
National Center for Health Statistics—see Department of Health and Human Services
National Highway Traffic Safety Administration—see Department of Transportation
National Institute of Child Health and Human Development—see Department of Health and Human Services
National Institute on Aging—see Department of Health and Human Services
National Institutes of Health—see Department of Health and Human Services
National Marine Fisheries Service—see Department of Commerce
National Oceanic and Atmospheric Administration—see Department of Commerce
National Park Service—see Department of the Interior
National Science Foundation—see National Science Foundation
Natural Resources Conservation Service—see Department of Agriculture
Occupational Safety and Health Administration—see Department of Labor
Office of the Assistant Secretary for Planning and Evaluation—see Department of Health and Human Services
Office of the Assistant Secretary for Policy Development and Research—see Department of Housing and Urban Development

Office of Environmental Information—see Environmental Protection Agency
Office of Health, Safety, and Security—see Department of Energy
Office of Immigration Statistics—see Department of Homeland Security
Office of Management and Budget—see Executive Office of the President
Office of Population Affairs—see Department of Health and Human Services
Office of Research, Evaluation, and Statistics—see Social Security Administration
Science Resources Statistics Division—see National Science Foundation, Directorate for Social, Behavioral, and Economic Sciences
Small Business Administration—see Small Business Administration
Social Security Administration—see Social Security Administration
Social Statistics Briefing Room—see Executive Office of the President, Federal Statistics Briefing Rooms
Statistics of Income Division—see Department of the Treasury
Substance Abuse and Mental Health Services Administration—see Department of Health and Human Services
United States Agency for International Development—see U.S. Agency for International Development
United States Fish and Wildlife Service—see Department of the Interior
United States Geological Survey—see Department of the Interior
United States Government Accountability Office—see U.S. Government Accountability Office

Appendix E

Prefaces to the First, Second, and Third Editions

FIRST EDITION

From time to time the Committee on National Statistics (CNSTAT) is asked for advice on what constitutes an effective federal statistical agency. For example, congressional staff raised the question as they were formulating legislation for a Bureau of Environmental Statistics, and the Secretary of the U.S. Department of Transportation asked CNSTAT for advice on establishing a new Bureau of Transportation Statistics, called for in the Intermodal Surface Transportation Efficiency Act of 1991. The National Research Council's Transportation Research Board had earlier turned to CNSTAT for information on common elements of the organization and responsibilities of federal statistical agencies for its study on strategic transportation data needs. Of interest in all of these requests are the fundamental characteristics that define a statistical agency and its operation.

Statistical agencies sometimes face situations that tax acceptable standards for professional behavior. Examples occur when policy makers, regulators, or enforcement officials seek access to data on individual respondents from a statistics agency or when policy interpretations are added to press releases announcing statistical data. Because the federal statistical system is highly decentralized, statistical agencies must operate under the policies and guidance of officials in many departments of government. Not all of these officials are knowledgeable about what is generally accepted as proper for a federal statistical agency, and issues involving judgments about conflicting objectives also arise.

In response to these situations, CNSTAT has prepared this "white paper" on principles and practices for a federal statistical agency. This paper brings together conclusions and recommendations made in many CNSTAT reports on specific agencies, programs, and topics, and it includes a discussion of what is meant by independence of a federal statistical agency and of the roles of research and analysis in a statistical agency. The commentary section contains supplementary information to further explain or illustrate the principles and practices.

In preparing this paper, CNSTAT and its staff solicited suggestions from many involved with federal statistical agencies. A draft of the paper was discussed by the heads of some federal statistical agencies at an open meeting of CNSTAT, and a draft was also discussed at a meeting of the Council of Professional Associations on Federal Statistics. The committee is grateful for the many suggestions and comments it received. When the report is published, CNSTAT plans to seek an even wider discussion of it at meetings of professional societies and to encourage reviews and commentaries. We hope that, in this way, the paper may evolve further and possibly influence legislation, regulations, and standards affecting federal statistical agencies.

As we were completing our work on this report, the Conference of European Statisticians drafted a resolution on the fundamental principles of official statistics in the region of the Economic Commission for Europe (ECE). Although the two documents were done independently, there is a large amount of agreement between them. We note particularly the emphasis the ECE resolution places on the need for independence for official statistics agencies (United Nations Statistical Commission and Economic Commission for Europe, 1991).

Although focused on federal statistical agencies, many of the principles and practices presented here also apply to statistical activities elsewhere, particularly to those in state and local government agencies and other statistical organizations. In addition, this paper and the ECE resolution may be useful to emerging democracies that seek to establish statistical organizations in their governments.

The principles and practices articulated here are statements of best practice rather than legal or scientific rules. They are based on experience rather than law or experiment. Some of them may need to be changed as laws change, society changes, or the practice of statistics changes. They are thus intended as guidelines, not prescriptions.

Burton H. Singer, *Chair*
Committee on National Statistics, 1992

NOTE: The ECE resolution was subsequently adopted by the Statistical Commission of the United Nations (U.N. Statistical Commission, 1994); see Appendix C.

SECOND EDITION

In 1992 the Committee on National Statistics (CNSTAT) issued a white paper on principles and practices for a federal statistical agency. The paper responded to requests from Congress and others for advice on what constitutes an effective statistical agency. It identified and commented on three basic principles: relevance to policy issues, credibility among data users, and trust among data providers. It also discussed 11 important practices, including a strong measure of independence and commitment to quality and professional practice (National Research Council, 1992).

The CNSTAT report has been used by federal statistical agencies to inform department officials, advisory committees, and others. It has also been used in a congressionally mandated study by the U.S. General Accounting Office (1995) to evaluate the performance of major statistical agencies and in a review of the federal statistical system by a former commissioner of the Bureau of Labor Statistics (Norwood, 1995). Its principles informed the establishment and later assessment of a new statistical agency, the Bureau of Transportation Statistics (see National Research Council, 1997b).

Eight years have passed since the white paper was first issued, and the committee decided that it would be useful to release a revised and updated version at this time. This second edition does not change the basic *principles* for federal statistical agencies, because the committee believes these principles are and will continue to be important guides for effective practice. The second edition does revise and expand the discussion of some of the *practices* that characterize an effective federal statistical agency and brings the discussion up to date with references to recent reports by the committee and others.

Driving the revisions is our recognition of the need for statistical agencies to keep up to date and to meet the challenges for their missions that are posed by such technological, social, and economic changes as the widespread use of the Internet for the dissemination and, increasingly, the collection of data, the heightened concern about safeguards for confidential information, and the information requirements of a changing economy. New and revised text addresses the reasons for establishing a federal statistical agency, the necessity for and characteristics of independence of a federal statistical agency, the need for continual development of more useful data, for example, by integrating data from multiple sources, practices for fair treatment of data providers, the role of the Internet in the release of data, and the need for effective coordination and cooperation among statistical agencies to ensure that policy makers and citizens receive data that are accurate, relevant, and timely for their needs.

We stress that the principles and practices for a federal statistical agency articulated here are guidelines, not prescriptions. We intend them to be helpful not only to the agencies, from whose experience we benefited in preparing this revised edition, but also to inform others of the characteristics of effective statistical agencies that can serve policy makers in the executive and legislative branches, other data users, and the public well.

John E. Rolph, *Chair*
Committee on National Statistics, 2001

THIRD EDITION

The Committee on National Statistics (CNSTAT) last revised its white paper on principles and practices for a federal statistical agency in 2001. First issued in 1992 on the committee's 20th anniversary, the white paper presents and comments on three basic principles for statistical agencies to carry out their mission effectively: relevance to policy issues, credibility among data users, and trust among data providers. The paper also discusses 11 important practices, including a strong measure of independence, a commitment to quality and professional practice, and an active program of methodological and substantive research.

The CNSTAT report has been widely cited and used by Congress and federal agencies. It has shaped legislation and executive actions to establish and evaluate statistical agencies, and agencies have used it to inform newly appointed department officials, advisory committees, and others about what constitutes an effective and credible statistical organization.

This third edition retains the outline and content of the second edition.

The changes and additions reflect new circumstances, such as new forms of threats to data confidentiality and individual privacy. This third edition also adds an appendix that documents legislation and regulations adopted since 2001 that importantly affect the operation of federal statistical agencies.

The principles and practices for a federal statistical agency articulated here remain guidelines, not prescriptions. We intend them to assist statistical agencies and to inform policy makers, data users, and others about the characteristics of statistical agencies that enable them to serve the public good.

William F. Eddy, *Chair*
Committee on National Statistics, 2005

COMMITTEE ON NATIONAL STATISTICS

The Committee on National Statistics (CNSTAT) was established in 1972 at the National Academies to improve the statistical methods and information on which public policy decisions are based. The committee carries out studies, workshops, and other activities to foster better measures and fuller understanding of the economy, the environment, public health, crime education, immigration, poverty, welfare, and other public policy issues. It also evaluates ongoing statistical programs and tracks the statistical policy and coordinating activities of the federal government, serving a unique role at the intersection of statistics and public policy. The committee's work is supported by a consortium of federal agencies through a National Science Foundation grant.